SpringerBriefs in Applied Sciences and Technology

For further volumes:
http://www.springer.com/series/8884

Theodore Hromadka · Robert Whitley

Foundations of the Complex Variable Boundary Element Method

 Springer

Theodore Hromadka
United States Military Academy
West Point, NY
USA

Robert Whitley
University of California
Irvine, CA
USA

ISSN 2191-530X ISSN 2191-5318 (electronic)
ISBN 978-3-319-05953-2 ISBN 978-3-319-05954-9 (eBook)
DOI 10.1007/978-3-319-05954-9
Springer Cham Heidelberg New York Dordrecht London

Library of Congress Control Number: 2014935988

Printed on acid-free paper

Springer is part of Springer Science+Business Media (www.springer.com)

To Laura
To Anne

Preface

This book introduces mathematical tools that are used in analyzing many physical problems. It has often been noted that, what at first appears to be an abstract generalization turns out to be not only useful but necessary for many applications.

To provide a focus, we establish what is needed to prove the main theorems of Chaps. 5 and 6, leaving unproved only two results from potential theory. The results of these two chapters improve on what we have previously published. These theorems provide easy methods for the numerical solution of the Dirichlet problem in two and three dimensions, which for concreteness we interpret as a problem in steady-state heat conduction. We refer to the literature for numerical applications, noting here only that such applications amount to a minimization over the parameters of the theorems.

If a reader is encouraged to learn more about Complex Variables or Linear Operators on Banach spaces, we will have accomplished our goal.

Theodore Hromadka
Robert Whitley

Acknowledgments

The authors are grateful for the help of several individuals and institutions that provided resources during the course of the research leading to the preparation of this manuscript. These institutions include University of California, Irvine, United States Military Academy at West Point, NY, and California State University, Fullerton. Special thanks go to our wives, Laura and Mikel, who encouraged the efforts of the authors throughout not only the current manuscript, but during the many prior research efforts. Special thanks are paid to the Cadets at the United States Military Academy at West Point who prepared numerous computer programs based on MatLAB and Mathematica, which demonstrate the efficacy of the mathematical principles presented in this manuscript. Further thanks are given to individuals at the Unites States Military Academy at West Point, including COL Michael Phillips, Ph.D., Head of the Department of Mathematical-Sciences, COL Steven Horton, Ph.D., Deputy Head, and the other faculty at the Department of Mathematical-Sciences, as well as Michael Barton for his editorial input.

Contents

Chapter 1
The Heat Equation

Abstract The problem of finding the steady state solution of the heat equation on the unit square, a basic example of the Dirichlet problem, is solved, and in the process ideas and theorems are developed, which are applicable to a broad range of mathematical problems. These important ideas include vectors spaces and linear operators, inner product spaces including R^N, Fourier series and their convergence, including Fejér's convergence theorem. The maximum principle is established for the case of the square.

Keywords Heat equation · Dirichlet problem · Vector spaces · Basis vectors · Linear operators · Inner product spaces · Cauchy-Schwarz · Triangle and Bessel's inequalities · Fourier series · Convergence of Fourier series.

A simple two-dimensional example of the Dirichlet problem can be stated on the unit square in the plane, with corners $(0, 0)$, $(1, 0)$, $(1, 1)$, and $(0, 1)$. Let a continuous function $g(x, y)$ be defined on the four edges of the square. The problem is to find a function $U(x, y)$ defined on the square which equals g on the edges of the square and is harmonic inside the square. Harmonic means that inside the square $U(x, y)$ has continuous second partial derivatives and there satisfies Laplace's equation:

$$\Delta U(x, y) = \frac{\partial^2 U(x, y)}{\partial x^2} + \frac{\partial^2 U(x, y)}{\partial y^2} = 0. \qquad (1.1)$$

The fundamental idea here is to consider the simplest example of a type of significant physical problem. In this case, the physical problem will be taken as a model for the conduction of heat, but other interpretations, for example in terms of fluid flow or gravitational potential, are possible.

The basic simplification is to reduce the problem from three dimensions to two dimensions. To do this, think of a very long rectangular metal bar, homogeneous so that the metal in any piece of the bar is the same as in any other piece, the long axis extending in the z-direction in three-dimensional space, with cross section a

T. Hromadka and R. Whitley, *Foundations of the Complex Variable Boundary Element Method*, SpringerBriefs in Applied Sciences and Technology, DOI: 10.1007/978-3-319-05954-9_1, © The Author(s) 2014

unit square when cut by places perpendicular to the z-direction. Apply heat to the sides of the bar in a way which does not vary with z; for example, hold the top of the bar at $1°$, the bottom at $-1°$, and the other two sides at $0°$. When the temperature in the bar has reached a steady state, U satisfies the steady state version of the heat equation:

$$\frac{\partial^2 U(x, y, z)}{\partial x^2} + \frac{\partial^2 U(x, y, z)}{\partial y^2} + \frac{\partial^2 U(x, y, z)}{\partial z^2} = 0 \qquad (1.2)$$

Because of the uniformity of the application of heat in the z-direction and the extreme length of the bar, there will be very little variation in heat in the z-direction in the middle of the bar, so that there the function U (approximately) satisfies Eq. (1.1).

The problem for the square needs to be formulated more precisely. As it stands, a function defined to be zero inside the square and $g(x, y)$ on the edges would be a solution, but would not represent the physical situation that is being modeled. In the physical situation, if a point (x, y) inside the square is close to a point (x_0, y_0) on the edge, the values of the temperature at (x, y), $U(x, y)$, should be close to the value of $g(x_0, y_0)$. This will occur if the solution $U(x, y)$ is also required to be continuous on the square plus its edges. Since the partial derivatives in (1.1) are required to exist and be continuous in the square and inside the square, $U(x, y)$ is continuous inside the square–the continuous extension to the edges is an additional requirement.

1.1 Separation of Variables

The problem of finding a function $U(x, y)$, satisfying (1.1), and taking on the values of a given function $g(x, y)$ on the edges of the square can be simplified. First, find $U_0(x, y)$ which agrees with a continuous function $g_0(x, 1), 0 \le x \le 1$ on the top of the square and is zero on the other three sides. If this can be done, then by a symmetrical argument three other solutions can be found which are each in turn equal to $g(x, y)$ on one of the other edges and are zero on the other three edges. Adding up four such solutions will give a continuous solution which satisfies (1.1) inside the square and is equal to the given function $g(x, y)$ on all the the four sides, the equality on each side being furnished by one of the four solutions.

When this simplification is made, the function $g_0(x)$, defined on the top of the square, may have values at $x = 0$ and $x = 1$ that conflict with the idea that the function $g_0(x) = g(x, 1), 0 \le x \le 1$ should be continuous on the edges of the square, $g_0(x)$ on the top, and zero on the other three sides–this is not possible if $g_0(0) \ne 0$ or $g_0(1) \ne 0$. This can be fixed, restating the problem by replacing $g_0(x)$ by $g_0(x) - [g_0(0) + x(g_0(1) - g(0))]$, which is now zero at $x = 0$ and $x = 1$. When this problem is solved for this new $g_0(x)$, to this solution $U_0(x, y)$ add the function $g_0(0) + x(g_0(1) - g(0))$, noting that this sum satisfies (1.1) and is equal to $g(x, y)$ on the top of the square.

Drop the subscripts on U and q, for implicity, and consider the asic problem of finding a continuous harmonic function $U(x, y)$ which agrees with a given continuous function $g(x)$ on the top of the square, $g(0) = 0$ and $g(1) = 0$, and is zero on the other three sides. To find a solution, we begin with The Method of Separation of Variables which looks for solutions to a simpler, but similar, problem in which $U(x, y)$ is the product of a function of x times a function of y, traditionally written as $X(x)Y(y)$. For such a function (1.1) becomes $X''Y + XY'' = 0$, the primes denoting the derivatives with respect to x and y. Write this as

$$\frac{X''}{X} = \frac{-Y''}{Y}$$

The key observation is that since the left-hand side of the equation depends only on x and the right-hand side depends only on y, it must be that both sides are constant, denoted by λ^2, so

$$\frac{X''}{X} = \frac{-Y''}{Y} = \lambda^2. \tag{1.3}$$

The general solution to (1.3) for the function X is $X = A \sin(\lambda x) + B \cos(\lambda x)$. Imposing the boundary condition $X(0) = 0$, the constant $B = 0$. From the condition $X(1) = 0$, possible values of λ are $\lambda_n = n\pi$, $n = 1, 2, \ldots$, the negative values of n giving no additional solutions since $\sin(-x) = -\sin(x)$. With each of the values of λ determined, the general solution for Y is $Y(y) = A e^{\lambda_n y} + B e^{-\lambda_n y}$. From the condition that $Y(0) = 0$, $Y(y)$ is a constant times $\sinh(\lambda_n y)$.

For each positive integer n, the function $A_n \sin(\lambda_n x) \sinh(\lambda_n y)$, A_n a constant, satisfies (1.1), therefore so does any (finite) sum of these functions. Putting aside possible technical difficulties for the moment, consider a possible solution which has the form of an infinite series

$$\sum_{n=1}^{\infty} A_n \sin(\lambda_n x) \sinh(\lambda_n y) \tag{1.4}$$

On the upper edge of the square, the boundary condition is:

$$\sum_{n=1}^{\infty} B_n \sin(n\pi x) = g(x) \tag{1.5}$$

with $B_n = A_n \sinh(n\pi)$. For this to be true, how should the coefficients B_n be chosen? The answer lies in elementary trigonometry. In the identity $\cos(\alpha + \beta) = \cos(\alpha)\cos(\beta) - \sin(\alpha)\sin(\beta)$, replace β by $-\beta$ for a second identity, subtract, and take $\alpha = n\pi x$ and $\beta = m\pi x$ to get

$$\sin(n\pi x)\sin(m\pi x) = \frac{1}{2}[\cos((n - m)\pi) - \cos((n + m)\pi)].$$

Multiple the series (1.5) by $\sin(m\pi x)$ and integrate over $[0, 1]$:

$$\sum_{n=1}^{\infty} B_n \int_0^1 \sin(n\pi x)\sin(m\pi x)dx = \int_0^1 \sin(m\pi x)g(x)\ dx.$$

Using the trig identity above, this reduces to a formula for the mth coefficient B_m in (1.5).

$$B_m \int_0^1 \sin^2(m\pi x)\ dx = \frac{1}{2}$$

and $B_m = 2\int_0^1 \sin(m\pi x)g(x)\ dx$. The question of when the resulting series converges to $g(x)$ and when the series for $U(x,y) = A_n \sin(\lambda_n x)\sinh(\lambda_n y)$ furnishes a solution to the steady state heat problem, will be addressed later. For now, at this point, several ideas, of great importance for mathematics and its applications, can be abstracted from this problem and so used with greater awareness in this and other problems.

1.2 Vector Spaces and Linear Operators

An example of a vector space is the representation of forces on an object in the plane. Each force is represented by an arrow with tail at the origin $(0, 0)$ and head at a point in the plane. Two vectors, one with head at (a, b) and the other with head at (c, d), can be added to get a vector which represents the resultant force due to the sum of these two forces. A graphical way of adding these two vectors is to translate, but not rotate, one so that its tail now lies at the head of the other vector; then its head is the head of the resultant vector (which has its tail at the origin). A simple diagram shows that this vector addition is the addition of the x and y components of the two vectors so that the resultant vector has head at $(a + c, b + d)$. Another aspect of representing vectors in the plane is that the magnitude of the force is given by the length of the vector; for a vector with tail at $(0, 0)$ and head at (a, b), Pythagoras's Theorem gives a force of $(a^2 + b^2)^{\frac{1}{2}}$. Thus, a force twice as large, but in the same direction, would be given by $2(a^2 + b^2)^{\frac{1}{2}} = ((2a)^2 + (2b)^2)^{\frac{1}{2}}$.

The extension to three dimensions is straightforward, but even more useful is the generalization to the vector space R^N, $N = 1, 2, \ldots$, where a vector consists of an N-tuple of real numbers $(a_1, a_2, \ldots, a_N)$. Addition is defined by:

$$(a_1, a_2, \ldots, a_N) + (b_1, b_2, \ldots, {}_N) = (a_1 + b_1, a_2 + b_2, \ldots, a_N + b_N),$$

and multiplication by a scalar, that is by a real number a, is defined by

$$a(a_1, a_2, \ldots, a_N) = (aa_1, aa_2, \ldots, aa_N).$$

The idea of a vector space, and related notions such as subspace and dimension, have found important applications throughout mathematics,where a vector space can consist of functions or more exotic objects. For example, solutions of the Dirichlet problem for the unit square can be constructed by multiplying simple solutions by constants and adding them to get a more complicated solution. In doing this, the solutions can be thought of as elements of a vector space. A vector space has two basic operations: adding vectors and multiplying vectors by real numbers, and these properties are directly abstracted from the properties of vectors in the plane R^2 or more generally R^N.

Definition 1 A real vector space is a collection V of objects called vectors. For two vectors $\mathbf{v}$ and $\mathbf{w}$, an addition is defined which gives another vector $\mathbf{v} + \mathbf{w}$ in V. This addition has the properties (of an Abelian group): For all $\mathbf{u}$, $\mathbf{v}$, and $\mathbf{w}$ in V: (1) $\mathbf{v} + \mathbf{w} = \mathbf{w} + \mathbf{v}$, (2) $\mathbf{v} + (\mathbf{w} + \mathbf{u}) = (\mathbf{v} + \mathbf{w}) + \mathbf{u}$, (3) There is an additive identity $\mathbf{0}$ with $\mathbf{v} + \mathbf{0} = \mathbf{v}$ for all $\mathbf{v}$, (4) Each $\mathbf{v}$ has an inverse $-\mathbf{v}$ with $\mathbf{v} + (-\mathbf{v}) = \mathbf{0}$.

For any vector $\mathbf{v}$ in V and real number a, a multiplication is defined which gives another vector $a\mathbf{v}$ in V. This multiplication has the properties: $\mathbf{v} = 1\mathbf{v}$ and, for a and b real, $a(b\mathbf{v}) = (ab)\mathbf{v}$. The vector addition and scalar multiplication are linked by: $a(\mathbf{v} + \mathbf{w}) = (a\mathbf{v}) + (a\mathbf{w})$, $(a + b)\mathbf{v} = (a\mathbf{v}) + (b\mathbf{v})$.

One further critical extension is to a complex vector space, which is defined exactly as above but with the scalars, a and b above, allowed to be complex numbers. It is common to use the term "vector space" to mean either a real vector space or a complex vector space, and to distinguish between the two only when special properties of the real or complex numbers are used. The complex version of R^N, C^N, is the collection of N-tuples $(c_1, c_2, \ldots, c_N)$ of complex numbers, with addition and multiplication by a complex scalar defined component-wise as for R^N.

Definition 2 A subspace M of a vector space V is a set of vectors from V which is itself a vector space under the operations inherited from V: if $\mathbf{v}$ and $\mathbf{w}$ are in M, so is $\mathbf{v} + \mathbf{w}$ and $a\mathbf{v}$ for any scalar a (i.e., a any real number, for a real vector space, or a any complex number for a complex vector space).

The span of a set of vectors S in V is the smallest subspace of V which contains S. This is the set of all linear combinations of vectors from S, i.e., all vectors which are finite sums of the form $a_1 \mathbf{s}_1 + \cdots + a_m \mathbf{s}_m$ for any $\mathbf{s}_j$ in S and any scalars a_j, and this subspace is denoted by $span(S)$.

The vectors in a collection of vectors $\mathcal{V}$ are linearly independent if a linear combination of vectors $\mathbf{v}_j$ from $\mathcal{V}$ is zero, that is a sum $a_1 \mathbf{v}_1 + a_2 \mathbf{v}_2 + \cdots + a_m \mathbf{v}_m = \mathbf{0}$, only if the coefficients $a_j = 0$ for all $j = 1, \ldots, m$.

A (finite) set of vectors in a vector space V, $\{\mathbf{v}_1, \mathbf{v}_2, \ldots, \mathbf{v}_m\}$, is a basis for V if they are linearly independent and any $\mathbf{v}$ in V can be written $\mathbf{v} = a_1 \mathbf{v}_1 + \cdots + a_m \mathbf{v}_m$. Then the vector space V is the span $V = span(\mathbf{v}_1, \ldots, \mathbf{v}_m)$, and the linear independence of the basis shows that for any $\mathbf{v}$ as above each coefficient a_j is uniquely determined since if there are two such expansions for $\mathbf{v}$, the difference is zero.

Example 1 The solutions $[\sin(\lambda_n x)\ \sinh(\lambda_n y)]$ for the Dirichlet problem for the square form a vector space; multiplication by a real number A_n gives the function

$A_n \sin(\lambda_n x) \sinh(\lambda_n y)$ and a (finite) sum is the function defined at x as the function whose value at x is the sum of the functions evaluated at x.

Example 2 Let V be the set of all polynomials in the variable x defined on the interval [0, 1]. For two polynomials **p** and **q**, addition is the addition of functions, i.e., the vector $\mathbf{p} + \mathbf{q}$ is the polynomial defined on [0, 1] by $(\mathbf{p} + \mathbf{q})(x) = \mathbf{p}(x) + \mathbf{q}(x)$ and multiplication by a scalar a by $(a\mathbf{p})(x) = a\mathbf{p}(x)$. It is easy to see that this defines either a real or a complex vector space, depending on the choice of scalars.

Let W be the subspace of V consisting of all polynomials of degree less than or equal to N. This subspace W has a basis consisting of the polynomials $1, \mathbf{x}, \mathbf{x}^2, \ldots, \mathbf{x}^N$. The linear independence of this basis follows since if $\mathbf{p}(x) = a_0 + a_1\mathbf{x} + \cdots + a_n\mathbf{x}^N = \mathbf{0}$, (**0** being the constant polynomial which is zero on [0, 1]), then $\mathbf{p}(x)$ is a polynomial of degree at most N which has more than N zeros, and so must be identically zero.

From calculus, the coefficient function for the jth coefficient a_j of a polynomial **p** is given by the jth derivative of **p**, divided by $j!$, evaluated at 0.

$$a_j = \frac{1}{j!} \frac{d^j \mathbf{p}}{dx^j}(0). \tag{1.6}$$

Theorem 1 *If V is a vector space with a basis $v_1, \ldots, v_n$ and with a basis $w_1, \ldots, w_m$, then $m = n$, and V is said to have dimension n.*

Proof Suppose $n < m$. Expand w_1 in terms of the v-basis: $w_1 = a_1 v_1 + \cdots + a_m v_m$. In this expansion, at least one $a_j \neq 0$. To simplify the notation, by re-indexing the v's if necessary, we can assume $a_1 \neq 0$. Then solving for v_1 shows that $V = span(w_1, v_2, \ldots, v_n)$. The vector w_2 is in V and so can be written $w_2 = a_1 w_1 + a_2 v_2 + \cdots + a_n v_v$. Since the w's are linearly independent, it cannot be the case that all the $a_j = 0$ for $j = 2, 3, \ldots, n$. By re-indexing if necessary, we can suppose that $a_2 \neq 0$. Solving for v_2 shows that $V = span(w_1, w_2, v_3, v_4, \ldots, v_n)$. In a finite number of steps, $V = span(w_1, w_2, \ldots, w_n)$. But then if $m > n$, w_{n+1} belongs to this span, which contradicts the linear independence of the w's. Q.E.D.

Example 3 If V has a basis $\mathbf{v}_1, \mathbf{v}_2, \ldots, \mathbf{v}_n$, then for each integer $j = 1, \ldots, m$, there is a function $\mathbf{v}_j^*$ defined for each $a_1 \mathbf{v}_1 + a_2 \mathbf{v}_2 + \cdots + a_m \mathbf{v}_m$ by

$$\mathbf{v}_j^*(\mathbf{v}) = a_j.$$

These coefficient functions have the properties that:

$$\mathbf{v}_j^*(\mathbf{v} + \mathbf{w}) = \mathbf{v}_j^*(\mathbf{v}) + \mathbf{v}_j^*(\mathbf{w}) \text{ for all } v \text{ and } w \text{ in } V, \tag{1.7}$$

$$\mathbf{v}_j^*(b\mathbf{v}) = b\mathbf{v}_j^*(\mathbf{v}) \text{ for all scalars } b. \tag{1.8}$$

Definition 3 A linear operator $T : V \rightarrow W$ mapping a vector space V into a vector space W is a function from V to W satisfying $T(a\mathbf{u} + b\mathbf{v}) = aT\mathbf{u} + bT\mathbf{v}$ for all vectors u and v in V and scalars a and b. A basic example displayed above, is the coefficient function $\mathbf{v}_j^*(\mathbf{v})$, the image space W being either the space of either real numbers or the complex numbers depending on whether V is a real or complex vector space. In this case, where W is the scalar field, as in example (3) the linear maps of V into W are called linear functionals.

Example 4 Solution to Eq. (1.1) can be obtained by summing constants times simple solutions because the two-dimensional Laplacian Δ is a linear operator defined on the vector space of those functions on the unit square, which have continuous second order partial derivatives.

Example 5 Let V be a vector space with basis $\mathbf{v}_1, \mathbf{v}_2, \ldots, \mathbf{v}_n$ and W a vector space with basis $\mathbf{w}_1, \mathbf{w}_2, \ldots, \mathbf{w}_m$. If $T : V \rightarrow W$ is a linear operator, each basis element $\mathbf{v}_j$ is mapped to an vector in W

$$T(\mathbf{v}_j) = a_{1j}\mathbf{w}_1 + a_{2j}\mathbf{w}_2 + \cdots + a_{mj}\mathbf{w}_m.$$

Then an arbitrary vector $\mathbf{v} = b_1\mathbf{v}_1 + b_2\mathbf{v}_2 + \cdots + b_m\mathbf{v}_m$ is mapped by T into the vector in W whose coefficient for $\mathbf{w}_k$ is

$$\sum_{j=1}^{n} a_{kj}b_j.$$

If the vector $\mathbf{v}$ is identified with the column vector whose jth component is b_j, then T can be identified with the n by m matrix $[a_{jk}]$, and the sum above is the formula for matrix multiplication.

Example 6 The vector space of polynomials in one variable has an important generalization to polynomials in N variables. The elegant notation used works like this: a typical term in a polynomial in the variables $x_1, x_2, \ldots, x_N$ has the form $x_1^{n_1} x_2^{n_2} \ldots x_N^{n_N}$, where n_1, n_2, and n_N are non-negative integers.

For $\alpha = (\alpha_1, \alpha_2, \ldots, \alpha_N)$, a vector where the components are non-negative integers, defining

$$\mathbf{x}^\alpha = x_1^{\alpha_1} x_2^{\alpha_2} \ldots x_N^{\alpha_N}$$

allows this term to be written in the compact form $\mathbf{x}^\alpha$. A polynomial in N variables is a finite sum of constants times the terms $\mathbf{x}^\alpha$; and the N-tuple α can be used to index these constants. So, for example, a general polynomial in three variables can be written

$$\sum C_\alpha \mathbf{x}^\alpha.$$

the sum being over some specified finite subset of α's, each α being a triple $\alpha = (n1, n2, n3)$ of non-negative integers n_1, n_2, n_3. With a little more notation, there is a neat generalization of the formula (1.6). For a function f of 3 variables written in component notation as $\mathbf{x} = (x_1, x_2, x_3)$, α as above, and

$$|\alpha| = \alpha_1 + \alpha_2 + \alpha_3,$$

define

$$D^\alpha f(\mathbf{x}) = \frac{\partial^{|\alpha|} f(x_1, x_2, x_3)}{\partial x_1^{\alpha_1} \partial x_2^{\alpha_2} \partial x_3^{\alpha_3}}$$

The final notation is $\alpha! = \alpha_1! \alpha_2! \alpha_3!$. Then for a polynomial $P(x_1, x_2, x_3)$ in three variables, partial differentiation shows that the coefficient C_α of the term $\mathbf{x}^\alpha$ is

$$C_\alpha = \frac{D^\alpha P}{\alpha!}(0, 0, 0),$$

which has the same form as (1.6). The corresponding formula for N variables follows from the same argument with a bit more notation. As has been noted above, this formula is a formula for the linear functional which is defined on the space of polynomials in N variable as the coefficient of the term $\mathbf{x}^\alpha$.

1.3 Inner Product Spaces

The Fourier Series in the solution of the heat equation has a structure that echoes the basic structure of the real number plane, that of an inner product space. This is a vector space on which is defined an inner product, (or scalar product), a map that takes each pair of vectors, $\mathbf{x}$ and $\mathbf{y}$ into a scalar $\langle \mathbf{x}, \mathbf{y} \rangle$ (alternate notation $(\mathbf{x}, \mathbf{y})$). For the case of a vector space over the scalar field of real numbers R, for vectors $\mathbf{x}, \mathbf{y}$, and $\mathbf{z}$, and α a real number, the inner product has the properties: (1) $\langle \alpha \mathbf{x}, \mathbf{y} \rangle = \alpha \langle \mathbf{x}, \mathbf{y} \rangle$, (2) $\langle \mathbf{x}, \mathbf{y} \rangle = \langle \mathbf{y}, \mathbf{x} \rangle$ (3) $\langle \mathbf{x} + \mathbf{y}, \mathbf{z} \rangle = \langle \mathbf{x}, \mathbf{z} \rangle + \langle \mathbf{y}, \mathbf{z} \rangle$, (4) $\langle \mathbf{x}, \mathbf{x} \rangle \geq 0$, and $\langle \mathbf{x}, \mathbf{x} \rangle = 0$ only if $\mathbf{x} = 0$. In R^N, for $\mathbf{x} = (x_1, x_2, \ldots, x_N)$, and $\mathbf{y} = (y_1, y_2, \ldots, y_N)$, the (usual) inner product is

$$\langle \mathbf{x}, \mathbf{y} \rangle = \sum_{j=1}^{N} x_j y_j$$

The length of $\mathbf{x}$ is denoted by $\|\mathbf{x}\|$ and is given by

$$\|\mathbf{x}\| = \left[\sum_{j=1}^{N} x_j^2 \right]^{1/2},$$

agreeing with the length given by the Pythagorean Theorem.

For a vector space over the complex numbers, the scalar α in property (1) is allowed to be a complex number, and property (2) must be modified to (2′) $\langle \mathbf{x}, \mathbf{y} \rangle = \overline{\langle \mathbf{y}, \mathbf{x} \rangle}$, where for the complex number $a + ib$, a and b real, $\overline{a + ib} = a - ib$ is the usual definition of the complex conjugate $a - ib$ of $a + ib$, consistent with the definition $\langle \mathbf{x}, \mathbf{y} \rangle = \sum_1^n x_j \overline{y_j}$. With this extension, for $\mathbf{x} = (x_1, \dots, x_N)$, an N-tuple of complex numbers,

$$\|\mathbf{x}\| = \left[\sum_{j=0}^{M} |x_j|^2 \right]^{1/2} ;$$

noting that the Pythagorean length of the complex number $x_j = a + ib$ is $|x_j| = |a^2 + b^2|^{1/2} = [(a + ib)(a - ib)]^{1/2}$. In the space R^N, regarded as an inner product space, the vectors $\mathbf{e}_1 = (1, 0, 0, \dots, 0)$, $\mathbf{e}_2 = (0, 1, 0, \dots, 0)$, and in general $\mathbf{e}_k$, having a 1 in the k-th component and zeros in the others, have the properties that $\langle \mathbf{e}_k, \mathbf{e}_m \rangle = 0$ for $k \neq m$, and $\langle \mathbf{e}_k, \mathbf{e}_k \rangle = 1$, and these properties show that any $\mathbf{x}$ in R^N can be written

$$\mathbf{x} = (x_1, x_2, \dots, x_N) = \sum_1^N x_n \mathbf{e}_n = \sum_1^N \langle \mathbf{x}, \mathbf{e}_n \rangle \mathbf{e}_n;$$

a similar expression holds for $\mathbf{x}$ in C^N.

The Fourier series which arose in the solution of the Dirichlet problem for the heat equation in two dimensions can be put in the same form, but with an infinite sum:

$$g(x) = \sum_1^{\infty} B_n \sin(n\pi x).$$

The coefficient $B_n = 2 \int_0^1 g(x) \sin(n\pi x)\, dx$, and the functions $\sin(\pi x)$, $\sin(2\pi x)$, ... play a role similar to the $\mathbf{e}_n$s. The integrals for the coefficients B_n suggest defining an inner product for functions f and g by

$$\langle f, g \rangle = \int_0^1 f(x) g(x)\, dx. \qquad (1.9)$$

Some technical conditions need to be met in order that the integral exists and defines a scalar product, but it is clear that there is a strong formal correspondence between the basis $\mathbf{e}_1, \dots, \mathbf{e}_N$ for R^N and the functions $\sin(\pi x)$, $\sin(2\pi x)$, ..., if the scalar product of two functions f and g on $[0, 1]$ is given by (1.9). In order to have functions of norm one, like the $\mathbf{e}_n$, the scaled functions $\sqrt{2}\sin(n\pi x)$ can be used and the formula for the coefficients B_n adjusted, or the inner product modified by multiplying it by $\frac{1}{2}$.

The following is a basic property of inner product spaces:

Lemma 1 *The Cauchy-Schwarz inequality.*

$$|\langle \mathbf{x}, \mathbf{y} \rangle| \leq \|\mathbf{x}\| \, \|\mathbf{y}\| \tag{1.10}$$

Proof For the case of a real inner product and scalar α, use the properties of an inner product to expand

$$\langle \mathbf{x} + \alpha \mathbf{y}, \mathbf{x} + \alpha \mathbf{y} \rangle = \|\mathbf{x}\|^2 + 2\alpha \langle \mathbf{x}, \mathbf{y} \rangle + \alpha^2 \|\mathbf{y}\|^2.$$

The value of α which minimizes this expression is $\alpha = -\frac{\langle \mathbf{x}, \mathbf{y} \rangle}{\|\mathbf{y}\|^2}$. Substitute this value for α and use the fact that the expression is non-negative to show that $\langle \mathbf{x}, \mathbf{y} \rangle \leq \|\mathbf{x}\| \, \|\mathbf{y}\|$; this also holds when $\mathbf{x}$ is replaced by $-\mathbf{x}$, which completes the proof for the case of real scalars.

For a complex-valued scalar product, the proof of (1.10) is similar. In the expansion of $\langle \mathbf{x} + \alpha \mathbf{y}, \mathbf{x} + \alpha \mathbf{y} \rangle$, replace $\mathbf{x}$ by $\beta \mathbf{x}$, where the complex number β is is chosen to have $|\beta| = 1$ and $\langle \beta \mathbf{x}, \mathbf{y} \rangle = |\langle \mathbf{x}, \mathbf{y} \rangle|$, and the proof, with α real, is then the same as for real scalars.   Q.E.D.

Corollary 1 *The Triangle Inequality.*
For $\mathbf{x}$ and $\mathbf{y}$ vectors in an inner product space

$$\|\mathbf{x} + \mathbf{y}\| \leq \|\mathbf{x}\| + \|\mathbf{y}\|$$

Proof Apply the Cauchy-Schwarz inequality to

$$\|\mathbf{x} + \mathbf{y}\|^2 = \|\mathbf{x}\|^2 + 2\,Re(\langle \mathbf{x}, \mathbf{y} \rangle) + \|\mathbf{y}\|^2.$$

Q.E.D.

Lemma 2 *Bessel's Inequality*
Let $\phi_1, \phi_2, \dots$ be a sequence of orthonormal vectors in an inner product space, i.e., $\langle \phi_n, \phi_m \rangle = 0$ if $n \neq m$ and $\langle \phi_n, \phi_n \rangle = 1$. For any $\mathbf{x}$ in the space,
$$\sum_1^\infty |\langle \mathbf{x}, \phi_n \rangle|^2 \leq \|\mathbf{x}\|^2.$$

Proof Let $a_n = \langle \mathbf{x}, \phi_n \rangle$. Then

$$0 \leq \left\| \mathbf{x} - \sum_{n=1}^N a_n \phi_n \right\|^2 = \|\mathbf{x}\|^2 - 2Re \sum_1^N \overline{a_n} \langle \mathbf{x}, \phi_n \rangle + \sum_1^N |a_n|^2 = \|\mathbf{x}\|^2 - \sum_1^N |a_n|^2$$

Thus $\sum_1^N |a_n|^2 \leq \|\mathbf{x}\|^2$, and let $N \to \infty$.   Q.E.D.

1.4 Convergence of Fourier Series

The validity of the solution of the heat equation on the unit square depends on the convergence of the Fourier series to the boundary function g, as well as the resulting solution being continuous on the square and satisfying the Laplace equation inside. The two convergence theorems here are therefore crucial, and we will see that the second result allows us to sidestep the technical problems involving an infinite series.

The notation for a Fourier series is simplified, and a more general series is produced, when the interval is changed from [0, 1] to $[-\pi, \pi]$, and an inner product defined by

$$\langle f, g \rangle = \frac{1}{\pi} \int_{-\pi}^{\pi} f(t)g(t) \, dt. \tag{1.11}$$

All the properties of an inner product are satisfied by this expression with f and g temporarily considered to be continuous functions defined on $[-\pi, \pi]$. With this interval and this inner product, Fourier series have the form:

$$g(x) = \frac{A_0}{2} + \sum_{n=1}^{\infty} A_n \cos(nx) + B_n \sin(nx). \tag{1.12}$$

The trigonometry identities now needed are easy consequences of

$$\cos(\alpha + \beta) = \cos(\alpha)\cos(\beta) - \sin(\alpha)\sin(\beta).$$

Change β to $-\beta$, use $\cos(-\beta) = \cos(\beta)$, i.e., cosine is an even function, and $\sin(-\beta) = -\sin(\beta)$, i.e., sine is an odd function, to obtain

$$\cos(\alpha - \beta) = \cos(\alpha)\cos(\beta) + \sin(\alpha)\sin(\beta).$$

Subtraction gives the identity which was used for the Fourier series on [0, 1],

$$\sin(\alpha)\sin(\beta) = \frac{1}{2}[\cos(\alpha - \beta) - \cos(\alpha + \beta)]$$

and

$$\cos(\alpha)\cos(\beta) = \frac{1}{2}[\cos(\alpha - \beta) + \cos(\alpha + \beta)].$$

From these two formulas it follows that

$$\langle \cos(nx), \cos(mx) \rangle = \frac{1}{\pi} \int_{\pi}^{\pi} \cos(nx)\cos(mx) \, dx = 0 \quad \text{for } n \neq m,$$

while $\langle \cos(nx), \cos(nx) \rangle = 1$. Similarly $\langle \sin(nx), \sin(mx) \rangle$ is zero for $n \neq m$ and one for $n = m$. The norms of $\cos(nx)$ and $\sin(nx)$ have been scaled to be one by the choice of the factor $\frac{1}{\pi}$ for the integral, instead of, say $\frac{1}{2\pi}$. That $\langle \cos(nx), \sin(mx) \rangle = 0$ is clear from the observation that the function $\cos(nx)\sin(mx)$ is an odd function and therefore its integral over $[-\pi, \pi]$ is zero. Thus an inner product of the Fourier series

(1.12) with the function $\cos(mx)$, if all problems of concerning the convergence of the series are ignored, the formula for the coefficient A_m and B_m are obtained:

$$A_m = \langle g(x), \cos(mx) \rangle \quad B_m = \langle g(x), \sin(mx) \rangle \tag{1.13}$$

The constant term is written $A_0/2$ so that the formula also holds for A_0 with $n = 0$.

The series (1.12), with coefficients given by the formulas above is the Fourier series of $g(x)$, and the convergence of that series to $g(x)$ is the subject of the next two theorems.

First we need to get ourselves into a little trouble. The functions $\sin(nx)$ and $\cos(mx)$ are defined for all real x, and in fact are periodic with period 2π, that is $\sin(n(x + 2\pi)) = \sin(nx)$ and $\cos(m(x + 2\pi)) = \cos(mx)$ and that holds for any real x, so you can picture the Fourier series repeating its values over the intervals $[-\pi, \pi], [\pi, 3\pi], [-3\pi, -\pi]$, and so on. It is convenient and will simplify calculations to extend the function $g(x)$, which was only defined on the interval $[-\pi, \pi]$, to be defined on all of R so as to be periodic with period 2π.

However, try extending the function $g(x) = x$, which is continuous before you extend it, to have period 2π. After it is given a 2π periodic extension it is no longer continuous but has jumps at each odd multiple of π, for example at the point π, the limit as you approach π from the left, $g(\pi^-) = \pi$, while the limit as you approach π from the right is $g(\pi^+) = -\pi$. This problem, in a hidden form, arose in the heat equation, and there a linear function $a + bx$ was subtracted from $g(x)$ to force the new function to be zero at the endpoints of $[0, 1]$. The best solution is to learn to live with functions that have jumps, say at most a finite number in any finite interval. We will do that, but then, using the Riemann integral in the definition of the inner product, the requirement that $\langle f, f \rangle = 0$ implies that f is zero can be violated. For example, the function defined on $[-\pi, \pi]$ which is zero except at the point 0 where $f(0) = 1$ has $\langle f, f \rangle = 0$ but f is not the zero function as it is not identically zero. This failure of one property of an inner product will be kept in mind until later results in a later chapter require a further discussion.

Consider the N-th partial sum of the Fourier series for a function g:

$$\sum_0^N A_n \cos(nx) + B_n \sin(nx)$$

$$= \frac{1}{\pi} \int_{-\pi}^{\pi} g(t) \left[\frac{1}{2} + \sum_1^N \cos(nt)\cos(nx) + \sin(nt)\sin(nx) \right] dt. \tag{1.14}$$

Using the trig identity for the cosine of the sum of two angles this can be written

$$\frac{1}{2} + \sum_1^N \cos(nt)\cos(nx) + \sin(nt)\sin(nx) = \frac{1}{2} + \sum_1^N \cos(n(x - t)). \tag{1.15}$$

Multiply (1.15) by $2\sin(x/2)$ and apply another trig identity to obtain the simplification:

$$2\sin(x/2)\left[\frac{1}{2} + \sum_1^N \cos(n(x-t))]\right] = \sin((N+1/2)(x-t)) \tag{1.16}$$

Lemma 3 *The Dirichlet Kernel. The N-th partial sum $S_N(x)$ for the Fourier series of g, a function of period 2π, can be written*

$$S_N(x) = \frac{1}{\pi}\int_{-\pi}^{\pi} g(t)D_N(x-t)dt = \frac{1}{\pi}\int_{-\pi}^{\pi} g(x+t)D_N(t)dt. \tag{1.17}$$

The Dirichlet kernel D_N is given by

$$D_N(t) = \frac{\sin[(N+1/2)(t)]}{2\sin(t/2)}, \tag{1.18}$$

and is a continuous function of period 2π.

Proof Take the limit of $D_N(t)$ as t tends to zero, using L'Hospital's rule, to show that $D_N(t)$ is continuous at zero, and therefore for all t. Although the numerator and denominator of $D_N(t)$ have period 4π, the quotient is easily seen to have period 2π. Make a change of variable in the first integral in (1.17), using $D_N(t) = D_N(-t)$, to obtain:

$$\frac{1}{\pi}\int_{-\pi-x}^{\pi-x} g(x+t)D_N(t)\,dt$$

Since the integrand has period 2π, the integral above is equal to the second integral in (1.17).   Q.E.D.

Definition 4 A function g defined for real x has a jump at the point x_0 if the limits from the left

$$g(x_0^-) = \lim_{\delta \to 0^+} g(x_0 - \delta)$$

and from the right

$$g(x_0^+) = \lim_{\delta \to 0^+} g(x_0 + \delta)$$

both exist, the limits being taken as $\delta > 0$ tends to zero, and $g(x_0^+) \neq g(x_0^-)$.

Theorem 2 *Let g be a function defined on the real line R which has period 2π and is continuous except perhaps for a finite number of jumps in $[-\pi, \pi]$. The Fourier series for g converges to $g(x_0)$ at any point x_0 where g is differentiable.*

Proof Integrating the sum in Eq. (1.16) shows that

$$\frac{1}{\pi} \int_{-\pi}^{\pi} D_N(t)\, dt = 1$$

from which

$$g(x_0) - S_N(x_0) = \frac{1}{\pi} \int_{-\pi}^{\pi} [g(x_0) - g(t + x_0)]\, D_N(t)\, dt. \tag{1.19}$$

Write the integrand as

$$\frac{g(x_0) - g(x_0 + t)}{t} \left[\frac{t}{2\sin(t/2)} \right] \sin[(n + 1/2)t].$$

The first term is continuous for all x, except for a possible finite number of jumps, and is continuous at x_0 where g is differentiable. The second term is continuous, at zero from an application of L'Hospital's rule showing that the limit is 1. So far,

$$h(t) = \frac{g(x_0) - g(x_0 + t)}{t} \left[\frac{t}{2\sin(t/2)} \right]$$

is continuous except possibly at a finite number of jumps. Since $\sin((N + 1/2)t) = \sin(Nt)\cos(t/2) + \cos(Nt)\sin(t/2)$, (1.19) equals

$$\frac{1}{\pi} \int_{-\pi}^{\pi} [h(t)\cos(t/2)] \sin(Nt)\, dt + \frac{1}{\pi} \int_{-\pi}^{\pi} [h(t)\sin(t/2)] \cos(Nt)\, dt.$$

Since the functions $h(t)\cos(t/2)$ and $h(t)\sin(t/2)$ are continuous except perhaps for a finite number of jumps, both have a square which is integrable over over $[-\pi, \pi]$ with a finite value, i.e., both have finite norms using the inner product (1.11). Thus, Bessel's inequality shows that the two integrals above tend to zero as N tends to infinity, which is to say that the partial sum $S_N(x_0)$ of the Fourier series for g converges to $g(x_0)$.

Q.E.D.

An interesting question, with a useful answer, is whether it is possible to generalize the notion of a limit of a sequence $s_1, s_2, \ldots$ so that the right answer is obtained if the limit exists, $s_n \to s$, and other sequences have a limit with this generalization even thought they do not converge. One such extension, Cesáro summability, considers the average $A_{n+1} = \frac{1}{n+1}(s_0 + s_1 + s_2 + \cdots + s_n)$. Suppose that the sequence $\{s_n\}$ converges to s. Since the s_n get close to s for large n, the sequence is bounded by, say, M. The convergence means that given an arbitrary $\epsilon > 0$, there is an N, with the property that $|s_n - s| < \epsilon$ for $n \geq N$. For $n > N$, estimate

$$|A_{n+1} - s| \le \frac{1}{n+1}\left[|s_0 - s| + \cdots |s_N - s)|\right] + \frac{1}{n+1}\left[|s_{N+1} - s| + \cdots + |s_n - s|\right]$$

which is bounded by $\frac{2(N+1)M}{n+1} + \frac{(n-N)\epsilon}{n+1}$, and is less than 2ϵ for large n. Hence, A_n converges to s. There are easy examples of sequences which do not converge themselves but whose associated average A_n converges–see exercises.

Theorem 3 *Let $S_n(x)$ be the n-th partial sum of the Fourier series for $g(x)$ on $[-\pi, \pi]$. The average*

$$A_{N+1}(x) = \frac{1}{N+1}\left[S_0(x) + S_1(x) + \cdots + S_N(x)\right] \tag{1.20}$$

can be written as the integral of g with respect to the Fejér kernel F_{N+1}

$$A_{N+1}(x) = \frac{1}{\pi}\int_{-\pi}^{\pi} g(x+t)F_{N+1}(t)\, dt \tag{1.21}$$

where

$$F_{N+1}(t) = \frac{\sin^2((N+1)(t/2)}{2(N+1)\sin^2(t/2)}. \tag{1.22}$$

Proof

$$\sum_0^N D_n(t) = \frac{1}{2\sin(t/2)}\sum_0^N \sin((n+1/2)t).$$

Multiply numerator and denominator by $\sin(t/2)$:

$$\frac{1}{2\sin^2(t/2)}\sum_0^N \sin(t/2)\sin((n+1/2)t).$$

The sum is

$$\sum_0^N \frac{1}{2}\cos(nt) - \cos((n+1)t) = \frac{1}{2}\left[\cos(0) - \cos(N+1)(t)\right].$$

and formula (1.22) follows, and from that (1.21). Q.E.D.

The kernel $F_N(t)$ has the properties:

(1) $F_N(t)$ is non-negative and has period 2π;
(2) $\int_{-\pi}^{\pi} F_N(t)\, dt = 1$. To see this, average the values of $\frac{1}{\pi}\int_{-\pi}^{\pi} D_n(t)$.
(3) For any $\delta > 0$, the integrals of $F_n(t)$ over $[\delta, \pi]$ and $[-\pi, -\delta]$ tend to zero as N tends to infinity. This follows from the fact that $\sin(t/2)$ increases on $[-\pi, \pi]$,

passing through zero at zero, and so $F_N(t)$ on $[-\pi, -\delta]$ and $[\delta, \pi]$ is bounded by $1/[N\sin^2(\delta/2)]$.

This means that for large N $F_N(t)$ has very little area outside a small interval about zero and a sharp peak at zero. If you have seen the Dirac delta function in physics, that is how $F_N(t)$ behaves as N tends to infinity, which is the idea in the proof below.

Theorem 4 *Suppose that $g(x)$ is a continuous function on the reals, periodic with period 2π. The average A_N in (1.20) of the partial sums of the Fourier series of g, which is to say the integral (1.21), converges uniformly to $g(x)$ on $[-\pi, \pi]$, i.e., given an $\epsilon > 0$, there is an N for which $|A_N(x) - g(x)| < \epsilon$ for all x in $[-\pi, \pi]$.*

Proof Two properties of a function g continuous on $[-\pi, \pi]$ will be used here, and proved in the next chapter. First, g is bounded on the interval, $|g(x)| \leq M$. Second, it is uniformly continuous there, so that given any $\epsilon > 0$, there is a $\delta > 0$ with $|g(x) - g(y)| < \epsilon$ for any $|x - y| \leq \delta$. (This δ works for all x in $[-\pi, \pi]$ and hence "uniformly," whereas in the case of g being continuous at each x in the interval, δ can depend on x.)

Proceeding as in the case of Theorem 2, write the difference between $g(x)$ and the average $A_N(x)$ (1.20) of the first N terms of the Fourier series for g in terms of the Fejér kernel

$$g(x) - A_N(x) = \frac{1}{\pi} \int_{-\pi}^{\pi} (g(x) - g(t+x))F_N(t)\, dt.$$

Write the integral as $\int_{-\pi}^{\delta} + \int_{-\delta}^{\delta} + \int_{\delta}^{\pi}$. The first integral is bounded by M times the integral of F_N over $[-\pi, -\delta]$ which, as noted above, can be made less than ϵ for large N. The same is true of the third integral. For the middle integral, since $|x-(x+t)| \leq \delta$, it is bounded by the integral of ϵ times the integral of F_N over a subinterval of $[-\pi, \pi]$, and it therefore bounded by ϵ. Thus for large N, $|g(x) - A_N(x)| < 3\epsilon$.   Q.E.D.

1.5 Maximum Principle

One way of attempting to solve the model problem on the square would be to attempt to find an a function $V(x, y)$ which approximately satisfied (1.1) inside the square and was approximately equal to $g(x, y)$ on the edges of the square. What will be done here instead is to approximate $g(x, y)$ on the edges using functions which are harmonic inside the square so that the approximation satisfies (1.1) exactly with no error inside the square–such methods are Boundary Element methods. But, supposing that this can be done and that we have obtained a function $V(x, y)$ which is harmonic inside the square and very close to $g(x, y)$ on the edges of the square, why should $V(x, y)$ also be close to the exact solution that we want, $U(x, y)$, on all of the square?

Start with the square D and a function $U(x, y)$ with continuous second partial derivatives and continuous on the square together with its edges, which satisfies the equation

$$\frac{\partial^2 U(x, y, z)}{\partial x^2} + \frac{\partial^2 U(x, y, z)}{\partial y^2} = G(x, y)$$

where G is a function $G(x, y) > 0$ in D and on the edges. Suppose that U has a maximum at a point (x_0, y_0) in the interior of D. Since (x_0, y_0) is in the interior of D there is an open circle inside of D containing that point, so if you fix $y = y_0$, there is an closed interval in x consisting of points of the form (x, y_0) containing (x_0, y_0) and on that interval function $U(x, y_0)$ has a maximum at the interior point x_0. By elementary Calculus, $\frac{\partial U}{\partial x}(x_0, y_0) = 0$ and $\frac{\partial^2 U}{\partial x^2}(x_0, y_0) \leq 0$. Apply the same argument to y_0 and add to get

$$\Delta U(x_0, y_0) = \frac{\partial^2 U(x_0, y_0)}{\partial x^2} + \frac{\partial^2 U(x_0, y_0)}{\partial y^2} \leq 0 < G(x_0, y_0)$$

a contradiction. However, we are interested in the Laplace equation $\Delta U = 0$. To get there, note that the function $H(x, y) = x^2 + y^2$ satisfies $\Delta H = 4$, and take $V = U + \epsilon H$. Then $\Delta V = 4\epsilon > 0$ for $\epsilon > 0$. By what we have shown above, V attains its maximum on the edges of D, using the fact that a continuous function on the closed square will attain its maximum there, which will be proven in the next chapter. Letting M denote the maximum of U on the boundary of D, for (x, y) in D

$$V(x, y) \leq M + \epsilon R^2$$

where D is contained in a circle of radius $R > \sqrt{2}$, which furnishes a bound on $x^2 + y^2$. Then since $U \leq V$ and $\epsilon > 0$ is arbitrary, $U \leq M$ in D.

Considering $-U$ shows that U also attains its minimum on the boundary, so that if we have found an approximate solution, which has boundary values which are within ϵ of the the values of a given boundary function, then this approximate solution is within ϵ of the exact solution in all of D.

The clever feature of this argument is looking at the case where ΔU is strictly positive in D in order to get the result for harmonic functions. Weinberger [1] interprets the result for the function $G(x, y) > 0$ as "This simply means that a strictly downward force at every point of a membrane cannot produce an upward bulge".

1.6 Exercises

1. In applying the method of separation of variables, Eq. (1.3) assumes that the constant, there called λ^2, is positive. Eliminate the possibility that the constant is zero or negative.
2. (Gram-Schmidt orthogonalization) Suppose that $v_1, v_2, \ldots, v_n$ are linearly independent vectors in an inner product space X. Consider the following inductive

construction: Let $\phi_1 = \frac{v_1}{\|v_1\|}$, making ϕ_1 a vector of norm one. Since v_2 does not belong to $sp(\phi_1)$,

$$w_2 = \phi_1 - \frac{\langle v_2, \phi_1 \rangle}{\|v_2\|^2} v_2$$

is not zero, and $\langle w_2, \phi_1 \rangle = 0$. Set $\phi_2 = \frac{w_2}{\|w_2\|}$. So far, ϕ_1, ϕ_2 are orthonormal, i.e., each has norm one and the two vectors are perpendicular. Continue this process, $w_3 = v_3 - [\langle v_3, \phi_1 \rangle \phi_1 + \langle v_3, \phi_2 \rangle \phi_2]$. Continue to get a set of n vectors $\phi_1, \phi_2, \ldots, \phi_n$ which are orthonormal and span the same subspace as the $v_1, \ldots, v_n$.

On X, define the operator $Pv = \sum_1^n \langle v, \phi_j \rangle \phi_j$. Show that P is a linear operator with $P^2 = P$. The operator P is the projection of X onto the subspace $V = span\{v_1, v_2, \ldots, v_n\}$ For any x in X, $x - Px$ is perpendicular to any v in V. Use this to calculate $\|x\|^2 = \|Px\|^2 + \|x - Px\|^2$. The operator $(I - P)x = x - Px$ is itself a projection onto all those vectors, which are perpendicular to all the vectors in V. This subspace is denoted by $V^\perp$ and is called the orthogonal complement of V.

In the vector space of all polynomials on the interval $[-1, 1]$, with inner product $\langle f, g \rangle = \int_0^1 f(x)g(x)dx$, and $v_1 = 1, v_2 = x, v_3 = x^2, \ldots$ the orthogonal polynomials obtained are the Legendre polynomials. There are numerous other orthogonal polynomials, obtained by using different inner products.

3. (a) Let f and g map the interval $[0, 1]$ into R^n and set $h(t) = \langle f(t), g(t) \rangle$. Show that $h'(t) = \langle f'(t), g(t) \rangle + \langle f(t), g'(t) \rangle$. (b) Suppose that $f(t)$ lies on the surface of a sphere of radius r with center $s_0 : \|f(t) - s_0\|^2 = r^2$. Show that the velocity $f'(t)$ on the sphere is perpendicular to the radius from center to that point, i.e., to $(f(t) - s_0)$. (c) Suppose that the point $f(t)$ on the surface of the sphere, taken to have center at zero, moves at constant speed, i.e., $\|f(t)\| = v$ is constant. Show that the velocity and acceleration vectors, $f'(t)$ and $f''(t)$, are perpendicular. For the case of $n = 2$, where the sphere is a circle in R^2, since $f'(t)$ is perpendicular to $f(t)$ and $f''(t)$ is perpendicular to $f'(t)$, $f''(t)$ is a scalar multiple of $f(t)$, say $f''(t) = af(t)$. Then

$$0 = \langle f'', f \rangle + \langle f', f' \rangle = \langle af, f \rangle + v^2 \text{ and } -v^2 = \langle af, f \rangle.$$

Thus $f''(t) = -\frac{v^2}{r^2} f(t)$. The interpretation is that when a point moves with constant speed on a circle of radius r, think of a stone tied to a string being swung in a circle, there is an acceleration that is directed towards the center of the circle and has magnitude $\frac{v^2}{r^2}$.

4. Let f be defined on $[-\pi, \pi]$. Show that f can be uniquely written as $f = f_e + f_o$, where f_e is an even function, i.e., $f_e(-x) = f_e(x)$, and f_o is an odd function, i.e., $f_o(-x) = -f_o(x)$. If the Fourier series for f is $\sum_0^\infty A_n \cos(nx) + \sum_1^\infty B_n \sin(nx)$, show that the first sum is the Fourier series for f_e and the second sum the Fourier series for f_o.

5. Give an example of a bounded sequence $\{s_n\}$ which does not converge but the sequence of averages $A_n = \frac{1}{n}[s_1 + \cdots + s_n]$ does converge. Give an example of a bounded sequence whose averages do not converge.

6. Compute the Fourier series of the function $f(x) = x$ on $[-\pi, \pi]$. By a choice of x, show that $\frac{\pi}{4} = 1 - \frac{1}{3} + \frac{1}{5} + \cdots$. Compute a few terms of the series. Estimate the error in approximating π by using N terms of this series. (This is an alternating series.)

7. Compute the Fourier series of the function $f(x) = |x|$ on $[-\pi, \pi]$. By a choice of x, show that $\frac{\pi^2}{8} = 1 + \frac{1}{3^2} + \frac{1}{5^2} + \cdots$. Compute a few terms of the series. Estimate the error in approximating π by the use of N terms of this series. (The integral test can be used here.)

8. Let f be continuous on $[-\pi, \pi]$, $f(-\pi) = f(\pi)$. Show that given $\epsilon > 0$, there is a trigonometric polynomial, i.e., a sum of the form $T_N(x) = \sum_0^N a_n \cos(nx) + \sum_1^N b_n \sin(nx)$ with $|f(x) - T_N(x)| < \epsilon$ for all x in $[-\pi, \pi]$.

9. Look up Gibbs phenomena for a discussion of the convergence of Fourier series at a jump [2].

10. Consider a string stretched between two points 0, and 1, and let $U(x, t)$ be the height of the string at the point x in $[0, 1]$ at the time $t \geq 0$. $[0, l]$. Begin with the string held in the shape $f(x)$, think of pulling and holding a guitar string with your finger at time $t = 0$. There is a good discussion in [1] of the physical approximations made which reduce the problem to the following, basically the values of $f(x)$ are not too large and the string thin and homogeneous. The function $U(x, t)$ satisfies the one-dimensional wave equation:

$$\frac{\partial^2}{t^2} U(x, t) = c^2 \frac{\partial^2 U(x, t)}{\partial x^2}$$

c a constant depending on the material of the string and $U(0, t) = U(l, t) = 0$, so that the ends of the string remain fixed, $U(x, 0) = f(x)$, $\frac{\partial U(x,0)}{\partial t} = 0$. This last equation denotes the fact that at time $t = 0$ no velocity has been imparted to the string. (a) Use separation of variables to find a solution of the form

$$U(x, t) = \sum_0^\infty b_n \sin(n\pi x/l) \cos(n\pi ct/l)$$

where $f(x) = U(x, 0)$.

(b) Use

$$\sin(n\pi x/l) \cos(n\pi c/l) = \frac{1}{2}[\sin(n\pi(x + ct)/l) - \sin(n\pi(x - ct)/l)]$$

to write the solution in the form

$$U(x, t) = \frac{1}{2}[f(x + ct) + f(x - ct)]$$

noting that the series implicitly extends f to have period l. Check that this solves the problem for f smooth enough. What does this solution indicate about how the shape of f is propagated along the string with time?

References

1. Weinberger HA (1965) First course in partial differential equations. Blaisdell Publishing Company, New York
2. Körner T (1988) Fourier analysis. Cambridge University Press, Cambridge

Chapter 2
Metric Spaces

Abstract Ideas are introduced to state and understand the Dirichlet problem in two, three, and N dimensions. These include open sets, boundaries, compact sets, and (uniformly) continuous functions. The concepts of completeness, sups and infs, compact sets, and continuous functions are used in the statement and proof of The Maximum Principle for the solution to the Dirichlet Problem in R^N and the Laplace equation in N variables.

Keywords Metric Space · R^N · Convergence · Geometric series · Continuous functions on a metric space · Open and closed sets · Compact sets · Compact sets in R^N · Sup · Inf · Completeness · The Maximum Principle

The combination of Theorem 4 and the maximum principle shows that there is an approximate solution to the heat equation for the unit square, given a continuous function g on the edges of the square, with error as small as desired. Restricting the continuous function g to the side with $0 \le x \le 1$, $y = 0$, it can be approximated to within a given $\epsilon > 0$ by the finite sum of of Theorem 4. If you add together the sums for each of the four sides, this sum $U_N(x, t)$ is harmonic inside the square and continuous on the edges because it is the finite sum of functions which have these properties. And, assuming that an exact solution $U(x, t)$ to this problem exists, the maximum principle shows that $|U_N(x, t) - U(x, t)| < \epsilon$ holds throughout the entire square. If, for example, you want to compute some isothermal lines, i.e., curved lines in the square where each point has the same constant temperature, you can compute them using $U_N(x, t)$ with the possible error as small as desired.

In obtaining these results, we have used the fact that the continuous function g is uniformly continuous and attains its maximum on [0, 1], and that $U_N(x, t)$ is continuous and also attains its maximum on the square and hence on the edge of the square. Further, in the proof of the Maximum Principle we have used general facts about the unit square, what is inside the square and what constitutes the edges of the square.

T. Hromadka and R. Whitley, *Foundations of the Complex Variable Boundary Element Method*, SpringerBriefs in Applied Sciences and Technology, DOI: 10.1007/978-3-319-05954-9_2, © The Author(s) 2014

This ideas need to be made precise so that domains more general than the square can be considered, in R^2 as well as in R^3. As is often the case, making the discussion more precise will open up wide areas of application. To discuss the accuracy of an (numerical) approximate solution of a problem a way of measuring the distance between solutions is required. A metric is such a measure.

Definition 5 A metric space is a set S on which is defined a measure of distance between points of S, a metric d, which is a map from pairs of elements of S to the non-negative real numbers with the properties, for x, y, and z in S: (1) $d(x, y) \geq 0$, and $d(x, y) = 0$ only if $x = y$, (2) $d(x, y) = d(y, x)$, and the triangle inequality (3) $d(x, z) \leq d(x, y) + d(y, z)$.

This definition abstracts the properties of the metric $d(x, y) = |x - y|$ given by the absolute value function on the real numbers.

Example 7 On R^N, for $\mathbf{x} = (x_1, x_2, \ldots, x_N)$ and $\mathbf{y} = (y_1, y_2, \ldots, y_N)$ the function

$$d_1(\mathbf{x}, \mathbf{y}) = \sum_1^N |x_j - y_j|.$$

satisfies the conditions for a metric.

Using the definition of the absolute value of a complex number $x + iy$, which is defined to be the length of the vector in the plane from $(0, 0)$ to (x, y), $|x + iy| = (x^2 + y^2)^{\frac{1}{2}}$, the formula above also gives a metric on C^N. In this case, to establish the triangle inequality requires the Pythagorean Theorem in the plane, which is a fact about a triangle.

Example 8 On R^N, for $\mathbf{x} = (x_1, x_2, \ldots, x_N)$ and $\mathbf{y} = (y_1, y_2, \ldots, y_N)$, a metric is given by
$$d_\infty(\mathbf{x}, \mathbf{y}) = \max \left(|x_j - y_j| : j = 1, \ldots, N \right).$$

and the same formula defines a metric on C^N with x_j and y_j in C.

Example 9 On any inner product space, $d(\mathbf{x}, \mathbf{y}) = \|\mathbf{x} - \mathbf{y}\|$ defines a metric, the triangle inequality for the metric is the triangle inequality for an inner product space. Examples include the usual Euclidean metric on R^N:

$$d_2(\mathbf{x}, \mathbf{y}) = \left[\sum_1^N |x_j - y_j|^2 \right]^{1/2}.$$

and the Euclidean metric on C^N, given by the same formula.

Definition 6 A sequence of points $x_1, x_2, \ldots, x_n, \ldots$ in a metric space X converges to a point x in X if $d(x_n, x) \to 0$ as $n \to \infty$.

For real numbers with the absolute value metric this is the usual definition of convergence.

An important example of a convergent sequence is given by the partial sums of a geometric series.

Lemma 4 *A geometric series has the form $\sum_0^\infty r^n$ where $-1 < r < 1$. The partial sums converge to $\frac{1}{1-r}$.*

Proof The Nth partial sum of the series is $S_N = \sum_0^N r^n$. Since $r S_N = r + r^2 + \cdots + r^{N+1}$, $S_N - r S_N = 1 - r^{N+1}$, and $S_N = \frac{1-r^{N+1}}{1-r}$. Let N tend to infinity.   Q.E.D.

It is a rare event that a sequence can be shown to converge to a specific value; in general the sequence defines the limit to which it converges. For example, suppose that $c_1, c_2, \ldots$ is a sequence where each c_n is one of the integers $0, 1, 2, 3, 4, 5, 6, 7, 8, 9$. Then

$$a = \frac{c_1}{10} + \frac{c_2}{10^2} + \frac{c_3}{10^3} + \cdots$$

is the decimal expansion of the number $a = 0.c_1 c_2 c_3 \ldots$ in the interval $[0, 1]$ How do we know that this represents a number in $[0, 1]$? What is meant by this infinite sum being equal to a real number is that the sequence of the partial sums $S_N = \sum_1^N \frac{c_j}{10^j}$ converges as N tends to infinity, the limiting value being indicated by $\sum_1^\infty \frac{c_n}{10^n}$. How can it be shown that this limit exists? Start by considering the series where all the $c_j = 1$, with partial sum $\sum_1^N \frac{1}{10^j}$ which by the lemma converges to $\frac{1}{9}$, since

$$S_N = \frac{1}{9}\left[1 - \frac{1}{10^N}\right].$$

What about the general number $0.c_1 c_2 \ldots$ in $[0, 1]$? Consider the difference between two partial sums $S_N - S_M$, $N > M$. Since each c_n satisfies $0 \le c_n \le 9$:

$$0 \le S_N - S_M = \sum_{M+1}^N \frac{c_j}{10^j} \le \frac{9}{10^M} \sum_1^{N-M} \frac{1}{10^j} < \frac{9}{10^M}.$$

From this, it follows that the difference $S_N - S_M$ tends to zero as N and M tend to infinity. So, intuitively, the terms of the sequence are getting close to each other for large N and M, so the sequence must be getting close to something.

Definition 7 A sequence $\{x_1, x_2, \ldots\}$ in a metric space with metric d is a Cauchy sequence if $d(x_n, x_m) \to 0$ as n, m tend to infinity. A metric space is complete if each Cauchy sequence converges; that is, for any Cauchy sequence $\{x_n\}$ there corresponds an element x in the space with $d(x_n, x) \to 0$ as $n \to \infty$.

It is a basic fact that the real numbers, with the usual absolute value metric, are complete. This is true because the real numbers are constructed to be complete, starting with the integers and with little beside mathematical induction as a tool.

The construction involves many details, and solves a mathematical problem which was open for over 2000 years, beginning when the ancient Greeks discovered that the rational numbers were not sufficient for geometry; one example being that the diagonal of a right triangle with sides each of length one is not a rational number—see exercises.

One consequence of the completeness of the real numbers is the existence of a real number for each decimal expansion. Generally whenever a numerical value is defined by a limiting process, completeness is usually required to show that this value exists.

Definition 8 Definition: Let X and Y be metric spaces and F a function mapping X to Y; $F: X \rightarrow Y$. The function F is continuous if whenever a sequence $\{x_n\}$ converges to x in X, the sequence $\{F(x_n)\}$ converges to $F(x)$.

One interpretation of continunity is that if a physical quantity is a continuous function F of x, then if you measure x accurately enough you will know $F(x)$ with as much accuracy as you need.

Another formulation of continunity comes from asking how closely the input value x must be known in order that the output be close to $F(x)$. In this form, if you want to know the value of $F(x)$ to within a certain error, traditionally denoted by an (arbitrary) value $\epsilon > 0$, it is required that the input value y be sufficiently close to x, this closeness is traditionally denoted by $\delta > 0$, where $|x - y| < \delta$. In spite of the fact that the appearance of two Greek letters in one definition makes it seem more complicated than the sequential definition above, the equivalent condition for continunity in the next lemma reflects an important physical property.

Lemma 5 *Let X be a metric space with metric d_x and Y another metric space with metric d_y. A function F mapping X to Y, $F: X \rightarrow Y$, is continuous at a point x_0 in X if and only if*

Given any $\epsilon > 0$, there is a number $\delta > 0$, depending on ϵ and x_0, with the property that if $d_x(x_0, x) < \delta$, then $d_y(F(x_0), F(x)) < \epsilon$.

Proof Suppose the $\delta - \epsilon$ condition holds. Let the sequence $x_n \rightarrow x_0$ in X, and let $\epsilon > 0$ be given. There is a $\delta > 0$ with the property that if $d_x(x, x_0) < \delta$ then $d_y(F(x), F(x_0)) < \epsilon$. Since $x_n \rightarrow x_0$, there is an integer N with $d_x(x_n, x_0) < \delta$ for $n \geq N$, and then $d_y(F(x_n), F(x_0)) < \epsilon$, showing that $F(x_n)$ converges to $F(x_0)$ and so that F is continuous at x_0.

Suppose that F is continuous at x_0. We show that the $\delta - \epsilon$ condition holds by assuming that it does not, i.e., for some $\epsilon_1 > 0$ an appropriate $\delta > 0$ cannot be found. This means that for each integer n and $\delta = 1/n$, there is a point x_n in X with $d_x(x_n, x_0) < 1/n$ but $d_y(F(x_n), F(x_0)) \geq \epsilon_1$. But then $x_n \rightarrow x_0$ yet $F(x_n)$ does not converge to $F(x_0)$, contrary to the continunity of F at x_0.   Q.E.D.

A metric space has two important related classes of sets. The applied heat equation problem of Chap. 1 was stated for the unit square, and the proof of the maximum principle for the square used the notion of points being inside the square and points being on the sides of the square. These ideas need to be extended to more general domains, and that is done in terms of two classes of sets in a metric space.

Definition 9 Let X be a metric space with metric d. (1) A subset A of X is open if for each point x in A, there is a $r > 0$ so that the sphere about x with radius r,

$$S(x, r) = \{y : d(y, x) < r\}$$

is contained in A: $S(x, r) \subset A$. (2) A subset B of X is closed if whenever a sequence of point $x_1, x_2, \ldots$ of points of B converges to x, this limit point x also belongs to B.

Lemma 6 *Let X be a metric space with metric d. A subset A of X is open if and only if the set B of all points of X which are not in A, written $B = X - A = A^c$ ("c" for the set complementary to A) is closed.*

Proof Let A be open and suppose that a sequence of points $x_1, x_2, \ldots$ from B converges to a point x. This point cannot be in A, for if it were, then because A is open there would be a sphere $S(x, r) \subset A$, and, by convergence the $\{x_n\}$ would be in A for large n, and these x_n would be both in A and the complementary set B, which is not possible.

Let B be closed, and let x be a point in A. We want to show that there is a sphere $S(x, r) \subset A$. Assume this is not true, then for each n the sphere $S(x, 1/n)$ is not contained in A and so there is a point x_n in this sphere and not in A therefore in B. But then the sequence of point $x_1, x_2, \ldots$ in B converges to x not in B, which cannot be.   Q.E.D.

In the metrics space R with the absolute value metric, the interval $(0, 1)$ is open, the interval $[0, 1]$ is closed, and the interval $(0, 1]$ is neither open nor closed.

For the unit square of the heat equation problem, the inside of the square is open and the edge of the square, as well as the entire square, is closed.

Definition 10 Let A be a set in a metric space X. The interior of A, written A^o is the largest open set contained in A. The closure of A, written $\overline{A}$, is the smallest closed set containing A.

Intervals in R, $(0, 1]^0 = (0, 1)$ and $\overline{(0, 1]} = [0, 1]$ The inside of the unit square is the interior of the square and the square (including its edges) is its own closure as it is itself closed.

The next idea needed in order to discuss harmonic functions on general domains corresponds to being given a continuous function g on the edge of the square. For an arbitrary set A the concept corresponding to the edge of the square is the boundary of A.

Definition 11 The boundary of a set A in a metric space X consists of those points x of X which have the property that any sphere $S(x, r)$ about x intersects both A ands its complement $X - A$.

In one dimension, if A is an interval—any of (a, b), $[a, b)$, or $[a, b]$—the boundary of A is the set consisting of the two points $\{a, b\}$.

The proof of the Maximal Principle relies on the fact that a function continuous on the closed unit square attains its maximum on the square.

Definition 12 For a set S of real numbers, the number $\sup(S)$, the supremum of S, is defined to be the least upper bound for the set S: so $M = \sup(S)$ if (1) M is an upper bound, i.e., all s in S satisfy $s \leq M$ and (2) it is the smallest upper bound, so that if $M' < M$, then there is at least one point s_0 in S for which $M' < s_0$.

In the older literature, the supremum of S was called the "least upper bound," which is much more descriptive than "sup," and was indicated by l.u.b.(S), pronounced "lub."

Example 10 $\sup((0, 1]) = 1$; $\sup((0, 1)) = 1$, in the the first case the supremum of the set $(0, 1]$ is attained in the sense that it belongs to the set $(0, 1]$ while this is not true for the set $(0, 1)$ having the same supremum.

The fact that a set of real numbers has a supremum is a consequence of the completeness of the reals. For an example where completeness is not available suppose you were considering sets of rational real numbers, and wanted to find in the rational numbers a supremum for any set with an upper bound. The set $S = \{x : x^2 < 2\}$ shows that this is not possible since there is no rational number which is a least upper bound for S. For any rational number which is an upper bound for S, there is a smaller rational number which is also an upper bound—just take a better rational approximation to the real number $\sqrt{2}$ which is the least upper bound for the set in the real numbers.

Lemma 7 *A set S of real numbers which has an upper bound, $s \leq M$ for all s in S, has a least upper bound.*

Proof Let B be the set of all real numbers which are upper bounds for S which is not empty since M belongs to B, and let A be all the real numbers which are not in B.

Begin an inductive procedure as follows. Choose a_1 in A. Since a_1 is not a bound for S there is a point s_1 in S larger than a_1. Set $b_1 = M$, and note

$$a_1 < s_1 \leq b_1.$$

and let $d = b_1 - a_1$. Consider the average value $c = \frac{a_1 + b_1}{2}$. If c is a bound for S, let $b_2 = c$, and let $a_2 = a_1$; but if c is not a bound for S, take $a_2 = c$ and $b_2 = b_1$. Then since a_2 is not a bound for S but b_2 is, there is a point s_2 in S, and

$$a_2 < s_2 \leq b_2$$

with $b_2 - a_2 = \frac{d}{2}$. After n steps of we have the points

$$a_n < s_n \leq b_n$$

with a_n not a bound for S, s_n in S, b_n a bound for S, with $b_n - a_n = \frac{d}{2^{n-1}}$. All three sequences are Cauchy and so all converge (to the same limit) b. For any s in S, $s \leq b_n$, and so $s \leq b$, showing that b is an upper bound for S. Further, if c is a real number less than b, since s_n converges to b, for large enough n, $s_n > c$, showing that b is indeed the smallest upper bound. Q.E.D.

There is a related concept for lower bounds for a set of real numbers.

Definition 13 For a set S of real numbers, the number $\inf(S)$, the infimum of S, is defined to be the greatest lower bound for the set S: so $N = \inf(S)$ if (1) N is an lower bound, i.e., all s in S satisfy $s \geq N$ and (2) it is the largest lower bound, so that if $N' > N$, then there is at least one point s_0 in S with $N' > s_0$.

In the older literature, the supremum of S was called the "greatest lower bound" and was indicated by g.l.b.(S), pronounced "glub."

Lemma 8 *A set S of real numbers which has a lower bound N, s $\geq$ N for all s in S, has a greatest lower bound.*

Proof Use $\inf(S) = -\sup(-S)$. Q.E.D.

One more idea is needed in order to work with general domains for the heat equation, we need the property that a continuous function on that domain will attain its maximum. If f is a continuous function mapping a domain D into the real numbers, with $M = \sup\{f(x) : x \in X\}$, then by the definition of the finite supremum M, for each $n = 1, 2, \ldots$, since $M - \frac{1}{n} < M$, and M is the smallest upper bound for f on D, there is a point x_n in D with $M - \frac{1}{n} < f(x_n) \leq M$; the second inequality following because M is a bound for all of the function values on D. With this in mind, you can see that if the sequence $x_1, x_2, \ldots$ converged to a point x in D, then we would have $f(x) = M$ and the supremum would be attained at this point x. In order to show that the supremum is attained, it would in fact be enough to have a subsequence converge; a subsequence of a sequence $\{x_n\}$ being another sequence formed from terms of the given sequence taken in the same order as the sequence. The property of compactness, defined below, has many applications.

Definition 14 Let $x_1, x_2, x_3, \ldots$ be a sequence of points (of a metric space X). By a subsequence of this sequence is meant a sequence formed using the points of the sequence taken in the same order: $x_{n1}, x_{n2}, x_{n3}, \ldots$ with $1 \leq n1 < n2 < n3 < \cdots$.

A set D in a metric space X is compact if any sequence of points $x_1, x_2, x_3, \ldots$ in D has a subsequence which converges to a point in D.

The following result has, more or less, been built-in to the definitions.

Theorem 5 *Let D be a set in a metric space X and f a continuous function mapping D into the real numbers. If D is compact, then f attains its maximum on D, i.e., there is a point x in D, with $f(x) = \sup\{f(y) : y \in D\}$.*

Since $\inf\{f(x) : x \in D\} = -\sup\{-f(x) : x \in D\}$, a continuous function also attains its infimum on a compact set.

In order to the apply results about compactness, we need to know which sets are compact. One property is clear: if a subsequence of points of D is going to converge to a point which is required to be in D, it seems that D must be closed. An additional property is needed, and in R^N what works is that the set be bounded.

Definition 15 A set D in R^N is bounded if it is contained in some sphere: i.e., $D \subset S(x_0, r)$, i.e., there is a point x_0 in R^N (which can be taken to be zero) with $d(x_0, x) < r$ for all x in D.

Theorem 6 *A subset of R^N is compact if and only if it is closed and bounded.*

Proof First consider a subset $D \subset R$, and a sequence of points $\{x_n\}$ from D. Since D is bounded it is contained in a sphere, which in R is an interval, say $D \subset [a, b]$, and then $|x_n - x_m| \le |b - a|$. There are an infinite number of integers $1 < n_1 < n_2 < \cdots$ for which either all the x_{n_j} are in $[a, \frac{a+b}{2}]$ or in $[\frac{a+b}{2}, b]$. It could be that both of these hold, but we know that at least one does. In any case, a subsequence is obtained satisfying: $|x_{n_j} - x_{m_j}| \le \frac{b-a}{2}$. To simplify the subscript notation, write $x_{n_j} = x_j^{(1)}$. The sequence $\{x_j^{(1)}\}$ has all its values in an interval of length $\frac{b-a}{2}$. Then there is a subsequence of this sequence which is contained in either the right-half or left-half of this new interval; let $\{x_j^{(2)}\}$ denote this subsequence and note that $|x_n^{(2)} - x_m^{(2)}| \le \frac{b-a}{2^2}$. Continuing this process for each integer k there is a subsequence $\{x_j^{(k)}\}$ of the previous subsequence for which $|x_n^{(k)} - x_m^{(k)}| \le \frac{b-a}{2^k}$.

Finally, the sequence with $y_n = x_n^{(n)}$ is a subsequence of the original sequence and satisfies $|y_n - y_m| \le \frac{b-a}{2^m}$ for $n \ge m$, and this sequence $\{y_n\}$ is a Cauchy sequence, which converges, since R is complete, to a point of D, since D is closed.

The proof for D a subset of R^N follows directly from the case for R. To simplify notation, consider R^2, and a sequence $\{(x_n, y_n)\}$ taken from the closed and bounded set D. The first coordinate sequence $\{x_n\}$ is bounded and so has a convergent subsequence $\{x_{n_j}\}$ from the result for R. The related second coordinates $\{y_{n_j}\}$ being a bounded sequence has a convergent subsequence. The sequence which has this as the second coordinate, has first coordinate which is also convergent since it is a subsequence of the convergent first coordinate sequence, and so this subsequence converges in R^2 since both its coordinates are convergent sequences; it converges to a point of D since D is closed. The proof for R^N proceeds along the same lines.

The idea of the proof is easy to visualize in R^2. A closed and bounded set D in R^2 is contained in a square. Let a sequence of points from D be given. Divide the square by two lines from the midpoints of opposite sides into four squares each one-quarter the size of the first square. There is a subsequence of the original sequence in at least one of the smaller squares. Keep up this process of subdividing the square to get subsequences which whose terms are in smaller and smaller squares and therefore closer and closer together. Then, as in the case of R, a subsequence can be obtained whose terms get closer and closer together, so that this sequence is Cauchy and so converges to a point in the closed set D. This is sometimes referred to as "Lion Hunting in the Sahara Desert". To find the lion (= the limit point of the subsequence) you enclose the desert in a square, divide the desert into quarters, pick the one the lion is in, divide the new square into quarters again, pick the one the lion is in, and continue until you have a square which is just big enough to hold the lion.

To show that D, a subset of R^N must be closed and bounded if it is compact, see the exercises.   Q.E.D.

The proof of Theorem 4 of Chap. 1 uses the fact below, a useful consequence of compactness.

Theorem 7 *Let K be a compact metric space with metric d and f a continuous function mapping K into the real numbers. Then f is uniformly continuous, i.e., given any $\epsilon > 0$, there is a $\delta > 0$ which works for all the x in K, i.e., for any x, if $d(x, y) < \delta$, then $|f(x) - f(y)| < \epsilon$.*

Proof Let f be as in the theorem and suppose that f is not uniformly continuous. This means that for some $\epsilon_0 > 0$ you cannot find a $\delta > 0$ that will work for all x; consequently for each positive integer n and $\delta = 1/n$, there are points x_n and y_n with $d(x_n, y_n) < 1/n$ but $|f(x_n) - f(y_n)| \geq \epsilon_0$. By the compactness of K, there is a subsequence $\{x'_n\}$ of $\{x_n\}$ that converges to a point x_0 in K. Since $d(y'_n, x_0) \leq d(x'_n, x_0) + d(x'_n, y'_n)$, the corresponding subsequence of the y_n converges to x_0 Passing to this subsequences and taking a limit shows that $|f(x_0) - f(x_0)| \geq \epsilon_0$, which cannot be true.   Q.E.D.

2.1 Maximum Principle

The concepts are now available for a statement of the Maximum Principle in a general setting.

Theorem 8 *Let D be a bounded open set in R^N with boundary Γ. Suppose that $U(x_1, x_2, \ldots, x_N)$ is a function with continuous second partial derivatives which satisfies the Laplace equation*

$$\Delta U(\mathbf{x}) = \sum_1^N \frac{\partial^2 U}{\partial x_j^2} = 0 \tag{2.1}$$

in D, and which is continuous on the closure of D. Then U attains its maximum (and minimum) on the boundary Γ of D.

Proof The proof is a simple adaptation of the proof of the the maximum principle for the square.   Q.E.D.

Note that the Maximum Principle is not an existence theorem, it indicates an important property of the solution assuming that there is a solution. The existence of a solution will be discussed later.

One consequence of the Maximum Principle is that there cannot be two different solutions U and V which satisfy the conditions of the above theorem and which are both equal to the same continuous function g on the boundary of K, for then $U - V$ satisfies the conditions of the theorem and is zero on the boundary of K, hence the maximum and the minimum of $U - V$ are both zero, which is to say that $U = V$.

2.2 Exercises

1. (Euclid) Show that $\sqrt{2}$ is not a rational number, i.e., it is not equal to the quotient of two integers n/m, $m \neq 0$. Suppose that $\sqrt{2} = n/m$. You can assume that n and m do not have a common factor. (If, say, $n = 3N$ and $m = 3M$, divide out the 3, and write $n/m = N/M$). Squaring, $2\,m^2 = n^2$. The square of an odd number is odd, so n must be even, $n = 2k$. Then $m^2 = 2k^2$, and m is also even; a contradiction.
2. Using the completeness of R^N, show that C^N is complete.
3. Show that a compact subset of R^N must be closed and bounded.
4. Show that a subset of C^N is compact if and only if it is closed and bounded.
5. Let X be a metric space with metric d. Define

$$d_0(x, y) = \frac{d(x, y)}{1 + d(x, y)}.$$

(i) Show that $d_0(x, y)$ is a metric, and that a subset of S is closed with respect to the metric d if and only if it is closed with respect to the metric d_0, and (X, d) and (X, d_0) also have the same open sets. (ii) Every subset A of X with the metric d_0 is bounded. (iii) Consequently, it is false in general that a subset of an arbitrary metric space is compact if it is closed and bounded. (iv) The subset A of the metric space (X, d) is totally bounded if every sequence of points of A has a Cauchy subsequence. Show that A is compact if and only it is closed and totally bounded.
6. Let f and g be continuous maps from a metric space X to the real numbers. Show that $f(x) + g(x)$, $f(x)g(x)$, and, if g is never zero, $\frac{1}{g(x)}$, are continuous.
7. Show that if K is a compact metric space and f a continuous map of K into another metric space, then f is uniformly continuous.
8. Give an example of a continuous real-valued function defined on $(0, 1)$ which is continuous but not uniformly continuous.
9. Show that $\sqrt{x}$ is uniformly continuous on $[0, 1]$.
10. Let f map the compact metric space X into the metric space Y. Show that the image of X under f, i.e., the set of all points y in Y for which there is an x in X with $y = f(x)$, is compact. For example, if K is a compact set in R^2, the projections onto the x-axis and y-axis are compact sets in R.

Chapter 3
Banach Spaces

Abstract Complete normed linear spaces are introduced, an example being the space $C(K)$ of continuous functions defined on a compact metric space K. For K the boundary of a bounded open set in R^N, this is the space to which the boundary functions for the Dirichlet problem belong. The space of polynomials of degree less than or equal to m, and the subspace of polynomials homogeneous of degree k, are finite dimensional spaces discussed here and which are used in the approximate solution of the Dirichlet problem. The Hahn-Banach Theorem is proved, and is a key ingredient in proving that the complex variable boundary element method will provide approximate solutions to the Dirichlet problem.

Keywords Normed linear space (n.l.s.) · Finite dimensional n.l.s · Bounded linear operators · Homogeneous and harmonic polynomials in N-variables · Complete n.l.s. = Banach space · Duality: The Riesz Representation Theorem · The Hahn-Banach Theorem.

In an inner product space, associated with each element $\mathbf{x}$ is a measure of its size or length, namely $\|\mathbf{x}\|$. The properties of this length are the basis for the more general notion of a normed linear space.

3.1 Normed Linear Spaces

Definition 16 A normed linear space X is a vector space, with scalars either the real or the complex numbers, on which is defined a mapping from the space to the real numbers, denoted, for $\mathbf{x}$ in X, by $\|\mathbf{x}\|$, called the norm of $\mathbf{x}$, which has the properties (1) $\|\mathbf{x}\| \geq 0$, and $\|\mathbf{x}\| = 0$ only when $\mathbf{x} = 0$, (2) For α a scalar, $\|\alpha\mathbf{x}\| = |\alpha|\|\mathbf{x}\|$, and (3) $\|\mathbf{x} + \mathbf{y}\| \leq \|\mathbf{x}\| + \|\mathbf{y}\|$, the triangle inequality.

The norm in a normed linear space will be used to make X a metric space with the metric $d(\mathbf{x}, \mathbf{y}) = \|\mathbf{x} - \mathbf{y}\|$, the properties of a metric following from those of a norm.

T. Hromadka and R. Whitley, *Foundations of the Complex Variable Boundary Element Method*, SpringerBriefs in Applied Sciences and Technology, DOI: 10.1007/978-3-319-05954-9_3, © The Author(s) 2014

Examples of norms on R^N or C^N have been indirectly given in the Chap. 2.

$$\|(x_1, x_2, \ldots, x_N)\|_1 = \sum_1^N |x_j|$$

$$\|(x_1, x_2, \ldots, x_N)\|_2 = \left[\sum_1^N |x_j|^2\right]^{1/2}$$

$$\|(x_1, x_2, \ldots, x_N)\|_\infty = \max\{|x_j|\}$$

Notation for these normed linear spaces will be $l^1(N)$, $l^2(N)$, and $l^\infty(N)$, respectively. Here, as usual, the same notation will be used for the two cases, one where the scalars, and therefore the coordinates $x_1, x_2, \ldots, x_N$, are real numbers, and the other where they are complex numbers.

There are infinite dimensional spaces which are direct generalizations of the finite dimensional example above.

The normed linear space l^1 consists of infinite sequences of scalars $\mathbf{x} = (x_1, x_2, \ldots)$ with the property that

$$\|\mathbf{x}\|_1 = \sum_1^\infty |x_j|$$

is finite. Recall that a series $\sum_1^\infty x_j$ with $\sum_1^\infty |x_j|$ finite is said to be absolute convergent; then the monotone non-decreasing sequence of the partial sums $\sum_1^N |x_j|$ converges (to their supremum $\sup_N \sum_1^N |x_j|$). By the completeness of the reals, this is equivalent to the partial sums $\sum_1^N |x_j|$ being a Cauchy sequence. Hence, from the inequality

$$\left|\sum_1^N x_j - \sum_1^M x_j\right| = \left|\sum_{M+1}^N x_j\right| \le \sum_{M+1}^N |x_j| = \sum_1^N |x_j| - \sum_1^M |x_j|$$

the partial sums $\sum_1^N x_j$ are a Cauchy sequence which then, by the completeness of the scalars, converges, i.e., the series $\sum_1^\infty x_j$ converges. With this in mind it is not difficult to see that $l^1 F$ is a normed linear space, with the definitions of vector addition and scalar multiplication analogous to those in $l^1(N)$.

The space l^2 is the collection of sequences $\mathbf{x}$ of scalars with the property that

$$\|\mathbf{x}\|_2 = \left[\sum_1^\infty |x_j|^2\right]^{1/2}$$

is finite, and, as above, with component addition and scalar multiplication. Apply the Cauchy-Schwarz inequality in $l^2(N)$ to the partial sums to see that

$$\sum_1^N |x_j \overline{y}_j| \le [\sum_1^N |x_j|^2]^{1/2}[\sum_1^N |y_j|^2]^{1/2} \le \|\mathbf{x}\|_2 \|\mathbf{y}\|_2$$

and the right-hand side is finite for $\mathbf{x}$ and $\mathbf{y}$ in l^2. It follows that the convergent series

$$\langle \mathbf{x}, \mathbf{y} \rangle = \sum_1^\infty x_j \overline{y_j}$$

defines an inner product which makes l^2 an inner product space with the given norm. Again, with the vector space operations for $\mathbf{x} = (x_1, x_2, \ldots)$, $\mathbf{y} = (y_1, y_2, \ldots)$, α and β scalars, being defined by

$$\alpha \mathbf{x} + \beta \mathbf{y} = (\alpha x_1 + \beta y_1, \alpha x_2 + \beta y_2, \ldots).$$

The space l^∞ is the collection of sequences $\mathbf{x}$ of scalars with the property that

$$\|\mathbf{x}\|_\infty = \sup\{|x_j|\}$$

is finite, so that l^∞ is the collection of all bounded sequences of scalars with the indicated norm, and is a normed linear space with this norm and component-wise addition and scalar multiplication as above.

Let K be a compact metric space and $C(K)$ denote the continuous functions mapping K into the scalars, either the real or complex numbers. This is a vector space under the natural definition that for f and g continuous on K and α and β scalars, $\alpha f + \beta g$ is the continuous function defined by $(\alpha f + \beta g)(x) = \alpha f(x) + \beta g(x)$. Bearing in mind that any continuous function f defined on K attains its maximum, a norm is defined on $C(K)$ by the finite value

$$\|f\| = \sup_{s \in K}\{|f(s)|\},$$

the supremum, as indicated, being taken over all the points s in K. With these definitions, $C(K)$ is an important example of a normed linear space.

The example of the functions f which are square integrable on $[0, 1]$, i.e., have $\int_0^1 |f(t)|^2 dt$ finite, is the Hilbert space $L^2[0, 1]$ with the inner product defined for f and g square integrable, as discussed in Sect. 1.3, by

$$\langle f, g \rangle \int_0^1 f(t)\overline{g(t)} \, dt$$

but this requires further comment.

The extension of the Riemann integral to the Lebesgue integral solves the problem of completeness, i.e., with the Lebesgue integral $L^2[0, 1]$ is complete, which is not so for the Riemann integral; there are technical details, but it will be enough to know that the extended integral equals the Riemann integral for all Riemann integrable functions. There is another issue. Consider the function f which is zero on $[0, 1]$ except at the point $\frac{1}{2}$ where it takes on the value 1. Then $\|f\| = [\int_0^1 |f(t)|^2 \, dt]^{1/2} = 0$, where the integral is the Riemann integral (or the Lebesgue integral), but f is not zero, and this violates the axiom that the norm of a function is zero only when the function is zero. The solution lies in the device of regarding two functions f and g which have $\int_0^1 |f(t) - g(t)|^2 dt = 0$, as "equivalent" and so as being the "same" function, but this will not be discussed here.

3.2 Bounded Linear Operators

The concept of a linear operator was discussed when solutions to the heat equation were found by forming linear combinations of solutions obtained by separation of variables. For normed linear spaces, the important linear operators are those which are continuous.

Definition 17 A linear operator $T: X \to Y$ mapping a normed linear space X into a normed linear space Y is bounded if there is a constant M for which

$$\|T\mathbf{x}\| \leq M\|\mathbf{x}\| \tag{3.1}$$

for all $\mathbf{x}$ in X. If T is bounded, then the smallest constant M for which (3.1) holds is denoted by $\|T\|$.

Theorem 9 *Suppose that X and Y are normed linear spaces and $T: X \to Y$, mapping X into Y is a linear operator. The operator T is continuous if and only if it is bounded.*

Proof Suppose that T is bounded, satisfying (3.1). Given $\epsilon > 0$ and $\mathbf{x}$, if $\|\mathbf{x} - \mathbf{y}\| < \delta = \frac{\epsilon}{M}$, then $\|T\mathbf{x} - T\mathbf{y}\| = \|T(\mathbf{x} - \mathbf{y})\| \leq M\|\mathbf{x} - \mathbf{y}\| \leq \epsilon$, and T is continuous. Conversely, if T is continuous at the zero vector in X, then given $\epsilon = 1$ there is a $\delta > 0$, so that $\|T(\mathbf{x}) - T(0)\| = \|T\mathbf{x}\| < 1$ for $\|\mathbf{x} - 0\| < \delta$. Then for an arbitrary $\mathbf{x}$ not zero, $\|\frac{\delta \mathbf{x}}{2\|\mathbf{x}\|}\| < \delta$, $\|T(\frac{\delta \mathbf{x}}{2\|\mathbf{x}\|})\| < 1$, so $\|T\mathbf{x}\| < \frac{2}{\delta}\|\mathbf{x}\|$ and T is bounded. Q.E.D.

Corollary 2 *Let X and Y be normed linear spaces, and $B(X, Y)$ the collection of all linear operators mapping X into Y. This is a vector space (over either the real or the complex numbers) when, for α and β scalers, and T and L linear operators, the linear operator $\alpha T + \beta L$ is defined by:*

$$(\alpha T + \beta L)(\mathbf{x}) = \alpha T\mathbf{x} + \beta L\mathbf{x},$$

and this vector space is a normed linear space under the operator norm defined above.

Proof The proof follows from the definitions.   Q.E.D.

3.3 Finite Dimensional Normed Linear Spaces

The norms in $l^1(N)$ and $l^2(N)$ are related by simple inequalities. Let $\mathbf{x} = (a_1, a_2, \ldots, a_N)$. Then

$$\|\mathbf{x}\|_1 = \sum_1^N |a_i| = \langle \mathbf{x}, (1, 1, \ldots, 1) \rangle \leq \|\mathbf{x}\|_2 \|(1, 1, \ldots, 1)\|_2 = \sqrt{N} \|\mathbf{x}\|_2 \quad (3.2)$$

and

$$\|\mathbf{x}\|_2 = [\sum_1^N |a_i|^2]^{1/2} \leq \left[\sum_1^N \|\mathbf{x}\|_1^2 \right]^{1/2} = \sqrt{N} \|\mathbf{x}\|_1. \quad (3.3)$$

For the linear operator T mapping $l^1(N)$ onto $l^2(N)$ by $T(a_1, a_2, \ldots, a_N) = (a_1, a_2, \ldots, a_N)$ (you would say this was the identity operator except that the range and domain are different normed linear spaces), inequality (3.2) shows that T is continuous and inequality (3.3) shows that the inverse operator, which is linear, is also continuous. These two norms are said to be equivalent and the two spaces are topologically equivalent. For example, a set K in one of the spaces is closed if and only if it is closed in the other space, bounded in one space if and only if it is bounded in the other space. Since closed and bounded sets are exactly the compact sets in $l^2(N)$ they are also the compact sets in $l^1(N)$, a fact which will be used in the next result.

Theorem 10 *Let X and Y be two N-dimensional normed linear spaces (over the same scalars). There is a one-to-one bounded linear operator T: X → Y mapping X onto Y which is continuous with a continuous inverse.*

Proof It will be enough to show that there is a one-to-one continuous operator T, with a continuous inverse, mapping $l^1(N)$ onto X. Then the same proof will give a continuous operator S with a continuous inverse, $S: l^1(N) \rightarrow Y$, and ST^{-1} will provide the operator taking X onto Y.

Let $\mathbf{e}_1, \mathbf{e}_2, \ldots, \mathbf{e}_N$ be a basis for X, and define $T: l^1(N) \rightarrow X$ by $T\mathbf{x} = T(a_1, a_2, \ldots, a_n) = a_1\mathbf{e}_1 + a_2\mathbf{e}_2 + \cdots a_N\mathbf{e}_N$. The operator T is bounded since $\| \sum_1^N a_i\mathbf{e}_i \| \leq \sum |a_i| \|\mathbf{e}_i\| \leq M \|(a_1, a_2, \ldots, a_N)\|_1$, with $M = \max\{\|\mathbf{e}_i\|\}$.

The operator T is one-to-one by the linear independence of the basis for X and is onto. It remains to show that T has a bounded inverse. The surface of the unit sphere in $l^1(N), B_1 = \{\mathbf{x} : \|\mathbf{x}\|_1 = 1\}$ is a compact subset of either R^N or C^N, depending on the scalars, because it is closed and bounded, and so the continuous function $\mathbf{x} \rightarrow \|T\mathbf{x}\|$

attains its minimum at a point $\mathbf{y}$ belonging to the set B_1. Since T is one-to-one and $\mathbf{y}$ is not zero, $\|T\mathbf{y}\| > 0$. Then for any $\mathbf{z} \neq 0$ in $l^1(N)$, $\|T(\frac{\mathbf{z}}{\|\mathbf{z}\|_1})\| \geq \|T\mathbf{y}\|_1$, or $\|T^{-1}(T\mathbf{z})\| \leq \frac{1}{\|T\mathbf{y}\|}\|T\mathbf{z}\|$, which is to say that T^{-1} is bounded.   Q.E.D.

One implication of Theorem 10 is that the unit ball in a finite dimensional normed linear space is compact, and so is the surface. This characterizes finite dimensional spaces.

Theorem 11 *If X is a normed linear space with the surface of the unit sphere, $B_1 = \{\mathbf{x} : \|\mathbf{x}\| = 1\}$, compact, then X is finite dimensional.*

Proof A set of points $\mathbf{x}_1, \mathbf{x}_2, \ldots, \mathbf{x}_m$ in B_1 is said to be a finite ϵ-net for a given $\epsilon > 0$, if for any $\mathbf{x}$ in B_1, there is a point $\mathbf{x}_j$ with $\|\mathbf{x} - \mathbf{x}_j\| < \epsilon$. Since B_1 is compact, for any $\epsilon > 0$, it has a finite ϵ-net, for if this is not true, then proceed as follows: Choose $\mathbf{x}_1$ in B_1. If the set consisting of the single point $\mathbf{x}_1$ is not an ϵ-net, then there is a point $\mathbf{x}_2$ in B_1 with $\|\mathbf{x}_2 - \mathbf{x}_1\| \geq \epsilon$. If the set $\{\mathbf{x}_1, \mathbf{x}_2\}$ is not an ϵ-net, then there is a point $\mathbf{x}_3$ in B_1 with $\|\mathbf{x}_3 - \mathbf{x}_1\| > \epsilon$ and $\|\mathbf{x}_3 - \mathbf{x}_2\| > \epsilon$. Continuing this process if a set $\mathbf{x}_1, \mathbf{x}_2, \ldots, \mathbf{x}_m$ of points of B_1 is not an ϵ-net , then there is a point $\mathbf{x}_{m+1}$ in B_1 with $\|\mathbf{x}_{m+1} - \mathbf{x}_j\| > \epsilon$ for $j = 1, 2, \ldots, m$. Then, assuming that at no stage do we get an ϵ-net, we wind up with an infinite sequence of points $\mathbf{x}_1, \mathbf{x}_2, \ldots$ with $\|\mathbf{x}_i - \mathbf{x}_j\| > \epsilon$ for $i \neq j$. It is clear that such a sequence is not a Cauchy sequence nor can it contain a Cauchy subsequence, which contradicts the compactness of B_1.

Supposing then that B_1 is compact, so that for a specific $0 < \epsilon_0 < 1$ there is an ϵ_0-net for B_1, namely $\mathbf{x}_1, \mathbf{x}_2, \ldots, \mathbf{x}_m$. I claim that $M = sp(\mathbf{x}_1, \ldots, \mathbf{x}_m)$, the subspace of X spanned by the vectors of the ϵ_0-net, is all of X, and hence X has dimension less than or equal to m. If instead M is a proper subspace of X, there is a point $\mathbf{y}_0$ in X with the distance from $\mathbf{y}_0$ to M positive, see definition 20, $d(\mathbf{y}_0, M) = d > 0$. (Note that M is closed since it is finite dimensional.) Since $d(\mathbf{y}_0, M)$, see definition 20, is defined to be the infimum of $\|\mathbf{y}_0 - \mathbf{x}\|$ taken over all $\mathbf{x}$ in X, there are points $\mathbf{z}_1, \mathbf{z}_2, \ldots$ in M with $\|\mathbf{y}_0 - \mathbf{z}_j\| \to d$ as $j \to \infty$. The sequence $\mathbf{z}_j$ is bounded in norm and belongs to the finite dimensional space M, and then, by compactness, has a subsequence converging to a point $\mathbf{z}$ of M with $\|\mathbf{y}_0 - \mathbf{z}\| = d$. Set $w = \frac{\mathbf{y}_0 - \mathbf{z}}{\|\mathbf{y}_0 - \mathbf{z}\|} = \frac{\mathbf{y}_0 - \mathbf{z}}{d}$. For any $\mathbf{x}$ in M,

$$\|\mathbf{w} - \mathbf{x}\| = \frac{\|\mathbf{y}_0 - \mathbf{z} - d\mathbf{x}\|}{d} \geq \frac{d}{d} = 1$$

since $\mathbf{z} + d\mathbf{x}$ belongs to M. This shows that the distance from $\mathbf{w}$ to any of the ϵ-net variables $\mathbf{x}_j$ is at least one, contradicting the property of being an ϵ-net with ϵ less than one.   Q.E.D.

Recall from Chap. 1 that a polynomial P in N variables can be written as the finite sum

$$P(\mathbf{x}) = \sum_{\alpha} C_\alpha \mathbf{x}^\alpha \tag{3.4}$$

where $\mathbf{x} = (x_1, x_2, \ldots, x_N)$, $\alpha = (\alpha_1, \alpha_2, \ldots, \alpha_N)$, the α_j all non-negative integers,

$$\mathbf{x}^\alpha = x_1^{\alpha_1} x_2^{\alpha_2} \ldots x_N^{\alpha_N}$$

Further, writing $|\alpha| = \alpha_1 + \cdots + \alpha_N$, the degree of $\mathbf{x}^\alpha$ is $|\alpha|$, and the degree of the polynomial given by (3.4) is the maximum of those numbers $|\alpha|$ corresponding to a non-zero coefficient C_α. The collection of all polynomials of degree less than or equal to m, so having the form

$$\sum_{|\alpha| \le m} C_\alpha \mathbf{x}^\alpha \tag{3.5}$$

is a vector space over the scalars, with the usual convention that the scalars are either the real numbers or the complex numbers. Setting

$$\|P\| = \sum_{|\alpha| \le m} |C_\alpha|$$

the polynomials of degree less than or equal to m is easily seen to be a normed linear space, and a Banach space since finite dimensional.

There are two subspaces of the polynomials of degree less than or equal to m which be used in the following.

First, a polynomial P is homogeneous of degree m if for all real numbers t

$$P(t\mathbf{x}) = t^m P(\mathbf{x})$$

where $t\mathbf{x} = (tx_1, tx_2, \ldots, tx_N)$ is scalar multiplication by scalar t, for $\mathbf{x}$ in R^N or C^N (depending on the scalars of the vector space of polynomials). Writing (3.5) as

$$P(\mathbf{x}) = \sum_m \sum_{|\alpha|=m} C_\alpha \mathbf{x}^m$$

shows that P can be written as the sum of the polynomials $\sum_{|\alpha|=m} C_\alpha \mathbf{x}^m$ each of which is homogeneous of degree m. A neat way of seeing that in the expansion of a polynomial P as a sum of homogeneous polynomials of degree m these polynomial are unique is to suppose that there are two such decompositions in terms of the polynomials P_m and Q_m homogeneous of degree m with

$$P(\mathbf{x}) = \sum_m P_m(\mathbf{x}) = \sum_m Q_m(\mathbf{x}), \tag{3.6}$$

and substitute $t\mathbf{x}$ for $\mathbf{x}$ in the sum to get

$$\sum_m t^m P_m(\mathbf{x}) = \sum_m t^m Q_m(\mathbf{x}) \tag{3.7}$$

Fixing the value of $\mathbf{x}$, Eq. (3.7) is a polynomial in t, written two ways with coefficients that must be equal (apply the derivative formula for the coefficients, i.e., use the fact that the powers of t are linearly independent in the vector space of polynomials over the reals).

In Chap. 1 harmonic functions on a domain were defined for two and three variables. It is easy to extend the definition to polynomials in N variables, saying that such a polynomial P is harmonic if

$$\Delta P(\mathbf{x}) = \frac{\partial^2 P(\mathbf{x})}{\partial x_1^2} + \cdots + \frac{\partial^2 P(\mathbf{x})}{\partial x_N^2} = 0 \qquad (3.8)$$

Using the fact that Δ is a linear operator and arguing as we did to show that the expansion in terms of homogeneous polynomials was unique, it follows that P is harmonic if and only if the homogeneous terms P_m in its expansion are themselves harmonic. In one direction, Eq. (3.6) shows that if each $P_m(\mathbf{x})$ is harmonic, then so is $P(\mathbf{x})$. In the other direction, if $P(\mathbf{x})$ is harmonic and t a real number,

$$0 = t^2 \Delta P(\mathbf{x}) = \Delta P(t\mathbf{x}) = \sum t^m \Delta P_m(\mathbf{x})$$

and as this is true for all real t, $\Delta P_m(\mathbf{x}) = 0$ for each $\mathbf{x}$ and all m. There is a norm on the space of harmonic polynomials that will be used in the proof of approximation theorems. Let D be an bounded open subset in R^N with boundary K, and let U be a function harmonic on D and continuous on the boundary K. Define

$$\|U\|_K = \sup\{|U(\mathbf{x})| : \mathbf{x} \text{ in } K\}$$

The interesting aspect of showing that this is a norm is in checking the condition that $\|U\| = 0$ implies that $U = 0$. This is true using the Maximum Principle, for if $U = 0$ on K, then its maximum and minimum on K, and therefore in D, are both zero and U itself is zero.

The sup norm expression above defines a norm on the space of harmonic polynomials, but if U is not harmonic, this is not necessarily true. For example, if D is the unit sphere in R^N, then the polynomial $U = 1 - [\|\mathbf{x}\|_2]^2$ is zero on the surface K, but U is not zero.

3.4 Completeness

Since the property of completeness is of basic importance, it is not surprising that those normed linear spaces which are complete (in the metric defined by the norm) are particularly useful.

Definition 18 A complete normed linear space is a Banach space.

The results of the last section show that all finite dimensional normed linear spaces are complete, since there is a one-to-one continuous map with a continuous inverse from $l^2(N)$ onto any N-dimensional normed linear space, and $l^2(N)$ is complete.

The space l^∞ is a Banach space. Since we have seen that this space of bounded scalar sequences $\{x_i\}$ with the sup norm $\|\mathbf{x}\|_\infty = \sup_i |x_i|$ is a normed linear space, we need only show completeness. Suppose that $\mathbf{x}^{(n)} = (x_1^{(n)}, x_2^{(n)}, \ldots)$ is a Cauchy sequence of vectors from l^∞, so that

$$\|\mathbf{x}^{(n)} - \mathbf{x}^{(m)}\|_\infty \to 0$$

as n and m tend to infinity. For any index j

$$|x_j^{(n)} - x_j^{(m)}| \le \|\mathbf{x}^{(n)} - \mathbf{x}^{(m)}\|_\infty \tag{3.9}$$

showing that the jth coordinate of the Cauchy sequence $\mathbf{x}^{(n)}$ is a Cauchy sequence of scalars and so converges to the scalar z_j. For a given $\epsilon > 0$, there is an N so that

$$\|\mathbf{x}^{(n)} - \mathbf{x}^{(m)}\|_\infty \le \epsilon$$

for n and m greater than N. Let m tend to infinity in (3.9) to see first that the sequence $\mathbf{z} = (z_1, z_2, \ldots)$ is bounded since its terms are each within ϵ of the bounded sequence $\mathbf{x}^{(N)}$, and so $\mathbf{z}$ belongs to l^∞. Next, letting m tend to infinity shows that $\mathbf{x}^{(n)}$ converges to $\mathbf{z}$ in l^∞.

Definition 19 A complete inner product space is a Hilbert space.

The space l^2 is a Hilbert space. Since l^2 is an inner product space, we need only show completeness. To do this, suppose that $\{\mathbf{x}^{(n)}\}$ is a Cauchy sequence of vectors in l^2, and write $\mathbf{x}^{(n)} = (x_1^n, x_2^n, \ldots)$. For each subscript i

$$|x_i^n - x_i^m| \le \|\mathbf{x}^{(n)} - \mathbf{x}^{(m)}\|$$

and it follows that for each fixed i, the sequence $\{x_i^{(n)}\}$ is a Cauchy sequence of scalars and so converges to a scalar z_i. Given $\epsilon > 0$ there is an N so that for any fixed M and $n \ge N$

$$\sum_{i=1}^{M} |x_i^n - x_i^{(m)}|^2 \le \|\mathbf{x}^{(n)} - \mathbf{x}^{(m)}\|^2 \le \epsilon^2$$

Let m tend to infinity in the above to obtain

$$\sum_{i=1}^{M} |x_i^{(n)} - z_i|^2 \le \epsilon^2 \tag{3.10}$$

for $n \ge N$ and any integer M. First, from the above and the triangle inequality

$$\Big[\sum_1^M |z_i|^2\Big]^{1/2} \le \Big[\sum_1^M |x_i^{(N)}|^2\Big]^{1/2} + \epsilon \le \|\mathbf{x}^{(N)}\|_2 + \epsilon$$

showing that $z = (z_1, z_2, \dots)$ is in l^2. Then, let M tend to infinity in (3.10) to see that the sequence $\mathbf{x}^{(n)}$ tends to the vector z in l^2.

Theorem 12 *Let K be a compact metric space and $C(K)$, the normed linear space of continuous function mapping K into the scalars with the supremum norm, is a Banach space.*

Proof We know that $C(K)$ is a normed linear space, and so only completeness needs to be shown.

Suppose that $f_1, f_2, \dots$ is a Cauchy sequence of functions in $C(K)$, given $\epsilon > 0$ there is an integer N with

$$\|f_n - f_m\| < \epsilon$$

for $n \ge N$ and $m \ge N$. For any x in K, since the norm is the sup norm, the sequence $f_1(x), f_2(x), f_3(x), \dots$ is a Cauchy sequence of scalars, which, since the scalars are complete, converges to a limit which depends on x and which we will denote by $f(x)$. This defines a function f mapping K into the scalars. There are two basic things to prove: (1) This function f is continuous, and so belongs to $C(K)$, and (2) The sequence $f_1, f_2, \dots$ converges to f in the metric of $C(K)$ induced by the norm.

Let $\epsilon > 0$ be given. Since the sequence of functions f_n is Cauchy, there is an N with $\|f_n - f_m\| < \epsilon$ for $n, m > N$. For any x in K,

$$|f_n(x) - f_m(x)| \le \|f_n - f_m\| \le \epsilon$$

for n and m greater than N. Letting m tend to infinity shows that

$$|f_N(z) - f(z)| \le \epsilon \tag{3.11}$$

for any z in K. For any x and y in K, the triangle inequality shows that

$$|f(x) - f(y)| \le |f(x) - f_N(x)| + |f_N(y) - f(y)| + |f_N(x) - f_N(y)| \tag{3.12}$$

and the expression above is bounded by $2\epsilon + |f_N(x) - f_N(y)|$. Since f_N is (uniformly) continuous on K, there is a $\delta > 0$ with the property that $d(x, y) < \delta$, implies $|f_N(x) - f_N(y)| < \epsilon$ and then $|f(x) - f(y)| < 3\epsilon$ and f is seen to be (uniformly) continuous on K. By (3.11), the sequence $\{f_n\}$ converges to f in the sup norm. Q.E.D.

Theorem 13 *Let X and Y be normed liner spaces. The normed linear space $B(X,Y)$ of all bounded linear operators mapping X to Y, taken with the operator norm, is complete if Y is complete.*

Proof Let T_n be a Cauchy sequence of operators in $B(X, Y)$. Given x in X,

$$\|T_n(x) - T_m(x)\| \le \|T_n - T_m\| \|x\|$$

which shows that the sequence $\{T_n(x)\}$ is Cauchy in Y, and so converges to a vector we denote by Tx. For x_1, x_2 in X and scalars α, β

$$T(\alpha x_1 + \beta x_2) = \lim T_n(\alpha x_1 + \beta x_2) = \alpha \lim T_n(x_1) + \beta \lim T_n x_2 = \alpha T(x_1) + \beta T(x_2)$$

showing that T is a linear operator. For $\epsilon = 1$, there is an N with $\|T_n - T_m\| \le 1$ for $n, m \ge N$, so that for $m \ge N$

$$\|T_m\| \le 1 + \|T_N\|.$$

Then $\|T(x)\| = \lim \|T_m(x)\| \le (1 + \|T_N\|)\|x\|$, showing that T is a bounded linear operator. Further, letting m tend to infinity in

$$\|T_m(x) - T_n(x)\| \le \|T_m - T_n\| \|x\|$$

shows that $\|T - T_n\| = \sup_{\|x\| \le 1} \|(T - T_n)(x)\|$ tends to zero as n tends to infinity, and the sequence $\{T_n\}$ converges to T in norm.   Q.E.D.

3.5 Duality

A normed linear space X has associated with it a Banach space which is, in a sense, at least as important as X; the dual, or conjugate space X^*. The space X^* is the space of bounded linear operators from X to the scalars, and is complete because the scalars, either the real or the complex numbers, are complete. The elements in X^* are known as bounded, or continuous, linear functionals.

Theorem 14 *The conjugate space of l^2 is l^2 in the following sense: Each continuous linear functional $\mathbf{x}^*$ on l^2 correspond to a unique element $\mathbf{y}$ in l^2 with $\mathbf{x}^*(\mathbf{x}) = \langle \mathbf{x}, \mathbf{y} \rangle$, and any $\mathbf{y}$ in l^2 gives rise to an bounded linear functional $\mathbf{x}^*$ as indicated. In this representation, $\|\mathbf{x}^*\| = \|\mathbf{y}\|$.*

Proof Let $\mathbf{e}_n$ be the vector in l^2 which is zero in all coordinates but the nth, where it is one: $\mathbf{e}_1 = (1, 0, 0, \ldots)$, $\mathbf{e}_2 = (0, 1, 0, 0, \ldots)$, etc. Any vector $\mathbf{x} = (x_1, x_2, \ldots)$ in l^2 can be written in terms of the $\{\mathbf{e}_n\}$

$$\mathbf{x} = \sum_1^\infty x_n \mathbf{e}_n$$

the $\{\mathbf{e}_n\}$ being an orthonormal basis for l^2. Recall that for a complex number α, the function $\mathrm{sgn}(\alpha)$, which extends the idea of the sign of a real number, is defined by $\mathrm{sgn}(\alpha) = \frac{\bar{\alpha}}{|\alpha|}$ for $\alpha \ne 0$ and is zero for $\alpha = 0$, so that $\mathrm{sgn}(\alpha)$ times α is $|\alpha|$. For a fixed N,

$$\mathbf{x}^* \left(\sum_1^N x_n \mathbf{e}_n \right) = \sum_1^N x_n \mathbf{x}^*(\mathbf{e}_n).$$

Let $y_n = \text{sgn}(\mathbf{x}^*(\mathbf{e}_n))|\mathbf{x}^*(\mathbf{e}_n)|$. For the choice $\mathbf{z}_N = \sum_1^N y_n \mathbf{e}_n$

$$\mathbf{x}^*(\mathbf{z}_N) = \sum_1^N |y_n|^2 \le \|\mathbf{x}^*\| \, \|\mathbf{z}_N\| = \|\mathbf{x}^*\| \left[\sum_1^N |y_n|^2 \right]^{1/2}$$

from which $\left[\sum_1^N |y_n|^2 \right]^{1/2} \le \|\mathbf{x}^*\|$. As N is arbitrary, $\mathbf{w} = \sum_1^\infty \mathbf{x}^*(\mathbf{e}_n)\mathbf{e}_n$ belongs to l^2, $\mathbf{x}^*(\mathbf{x}) = \langle \mathbf{x}, \mathbf{w} \rangle$, and the norm of $\mathbf{x}^*$ is given by $\|\mathbf{w}\|$.   Q.E.D.

Theorem 15 *The conjugate space of l^1 is l^∞ in the sense that each continuous linear functional $\mathbf{x}^*$ on l^1 corresponds to a bounded sequence $(a_1, a_2, \dots)$. For $\mathbf{x} = (x_1, x_2, \dots)$ in l^1,*

$$\mathbf{x}^*(\mathbf{x}) = \sum_1^\infty a_n x_n$$

and $\|\mathbf{x}^*\| = \sup|a_n| = \|(a_1, a_2, \dots)\|_\infty$.

Proof Let $\mathbf{e}_n$ be the vector in l^1 which is zero in all coordinates except for the nth, where it is one. Any vector $\mathbf{x} = (x_1, x_2, \dots)$ in l^1 can be written in terms of the $\{\mathbf{e}_n\}$

$$\mathbf{x} = \sum_1^\infty x_n \mathbf{e}_n, \tag{3.13}$$

and $\|\mathbf{x}\|_1 = \sum_1^\infty |x_n|$. Since the series (3.13) converges in norm to $\mathbf{x}$ and $\mathbf{x}^*$ is continuous,

$$\mathbf{x}^*(\mathbf{x}) = \sum_1^\infty \mathbf{x}^*(\mathbf{e}_n) x_n.$$

Set $a_n = \mathbf{x}^*(\mathbf{e}_n)$. Since $|a_n| \le \|\mathbf{x}^*\|$, the sequence $\mathbf{y} = (a_1, a_2, \dots)$ belongs to l^∞, and $|\sum_1^\infty a_n x_n| \le \|\mathbf{x}\|_1 \|\mathbf{y}\|_\infty$. Because the vector $\mathbf{e}_n$ has norm one in l^1, it follows that the norm of $\mathbf{x}^*$ is the sup norm of $\mathbf{y}$.   Q.E.D

3.6 Riesz Representation Theorem

In the last section it was shown that any continuous linear function on l^2 was given by an inner product. As will be shown here, this is true for any Hilbert space, and this fact is the Riesz Representation Theorem.

Definition 20 If N is a subspace of a normed linear space X, in particular a subspace of a Hilbert space, the distance $d(\mathbf{x}, N)$ of a point $\mathbf{x}$ of X to the subspace N is defined by:

$$d(\mathbf{x}, N) = \inf\{\|\mathbf{x} - \mathbf{y}\| : \mathbf{y} \text{ in } N\}$$

where, as indicated, the infimum is taken over all $\mathbf{y}$ in N.

Lemma 9 *Let N be a closed subspace of a Hilbert space X and $\mathbf{x}$ a point of X. There is a unique point $\mathbf{z}$ belonging to N with $d(\mathbf{x}, N) = \|\mathbf{x} - \mathbf{z}\|$.*

Proof By the definition of the distance, there are points $\mathbf{z}_n$ in N with $\|\mathbf{x} - \mathbf{z}_n\| \to d(\mathbf{x}, N)$. A property of the norm in a Hilbert space, known as the parallelogram law, which can be seen by expanding the norms on the left-hand side as inner products, is

$$\|\mathbf{x} + \mathbf{y}\|^2 + \|\mathbf{x} - \mathbf{y}\|^2 = 2\|\mathbf{x}\|^2 + 2\|\mathbf{y}\|^2.$$

Apply this to the vectors $\mathbf{x} + \mathbf{x}_n$ and $\mathbf{x} - \mathbf{x}_m$ to see that

$$4\left\|\mathbf{x} - \frac{\mathbf{x}_n - \mathbf{x}_m}{2}\right\|^2 + \|\mathbf{x}_n - \mathbf{x}_m\|^2 = 2\|\mathbf{x} - \mathbf{x}_n\|^2 + 2\|\mathbf{x} - \mathbf{x}_m\|^2.$$

As the right-hand side converges to $4d^2$, and the first term of the left-hand side is at least $4d^2$, it follows that the sequence $\{\mathbf{x}_n\}$ is Cauchy and so converges to a vector $\mathbf{z}$ in N, and this vector has $\|\mathbf{x} - \mathbf{z}\| = d$.

To see the uniqueness, suppose that there are two points $\mathbf{z}_1$ and $\mathbf{z}_2$ in N with $d = \|\mathbf{x} - \mathbf{z}_1\| = \|\mathbf{x} - \mathbf{z}_2\|$. From the parallelogram law

$$4\left\|\mathbf{x} - \frac{\mathbf{z}_1 + \mathbf{z}_2}{2}\right\|^2 + \|\mathbf{z}_1 - \mathbf{z}_2\|^2 = 4d^2.$$

Since $\frac{\mathbf{z}_1 + \mathbf{z}_2}{2}$ belongs to N, the first term is at least $4d^2$, and so $\mathbf{z}_1 = \mathbf{z}_2$.   Q.E.D.

Definition 21 For two subspaces N and M of a vector space V, we write $V = M + N$ if any v in V can be written as $v = x + y$, X in N and y in M. If M and N have only the zero vector in common, we write $V = M \oplus N$; in this case the representation of v in V as $v = x + y$, x in M and y in N, is unique, for if $v = x_1 + y_1$ and $v = x_2 + y_2$, then $x_1 - x_2 = y_2 - y_1$ is in both N and M and so is zero.

For a subspace N of a Hilbert space X, the closed subspace $N^\perp$ is defined to consist of all the vectors $\mathbf{x}$ which are perpendicular to all the vectors in N.

Lemma 10 *If N is a closed subspace of Hilbert space X,*

$$X = N \oplus N^\perp.$$

Proof Given $\mathbf{x}$ not in N, there is a unique $\mathbf{z}$ in N with $d(\mathbf{x} - \mathbf{z}, N) = \|\mathbf{x} - \mathbf{z}\| = d(\mathbf{x}, N)$. Set $\mathbf{w} = \mathbf{x} - \mathbf{z}$. The vector $\mathbf{w}$ is perpendicular to N. For if not, there is a vector $\mathbf{y}$ in N with $\langle \mathbf{w}, \mathbf{y} \rangle \neq 0$. Expand

$$\|\mathbf{w} + \alpha\mathbf{y}\|^2 = \|\mathbf{w}\|^2 + 2Re\,(\alpha\langle\mathbf{y},\,\mathbf{w}\rangle) + |\alpha|^2\|\mathbf{y}\|^2.$$

Choose the scalar α so that $\alpha\langle\mathbf{y},\,\mathbf{w}\rangle$ is real and negative and small enough that the middle term is in absolute value larger than the last term. For real scalars, to get the sign right just requires multiplying $\mathbf{y}$ by plus or minus one, and then multiplying by a small positive real number α. For complex scalars this requires multiplying $\mathbf{y}$ by $-\text{sgn}(\langle\mathbf{y},\,\mathbf{w}\rangle)$, where for a complex number β, $\text{sgn}(\beta) = \frac{\beta}{|\beta|}$ is the complex number extension of the sign of a real number, having the property that it is a complex number with $\beta\,\text{sgn}(\beta){=}|\beta|$. This done, the resulting scalar multiple of $\mathbf{y}$ still belongs to N and the middle term is negative and with absolute value larger than the last term, so it follows that

$$d(\mathbf{w},\,N)^2 \le \|\mathbf{w} + \alpha\mathbf{y}\|^2 < \|\mathbf{w}\|^2 = d(\mathbf{w},\,N)^2,$$

a contradiction. So $\mathbf{x} = \mathbf{w} + \mathbf{z}$, the first term is in $N^\perp$, the second in N, and clearly $N^\perp$ and N intersect only in the zero vector.   Q.E.D.

Theorem 16 *The Riesz Representation Theorem. For a Hilbert space X, any continuous linear functional* $\mathbf{x}^*$ *on X has the form* $\mathbf{x}^*(\mathbf{x}) = \langle\mathbf{x},\,\mathbf{y}\rangle$ *for a unique vector* $\mathbf{y}$ *in X. And any such inner product gives a bounded linear functional on X with* $\|\mathbf{x}^*\| = \|\mathbf{y}\|$.

Proof Let N be the set of all vectors $\mathbf{x}$ with $\mathbf{x}^*(\mathbf{x}) = 0$; i.e., N is a closed subspace known as the null manifold of the continuous linear functional $\mathbf{x}^*$. By Lemma 2 there is a vector $\mathbf{x}_0$ which is perpendicular to N; by dividing $\mathbf{x}_0$ by $\|\mathbf{x}_0\|$ we may suppose that $\|\mathbf{x}_0\| = 1$. If $\mathbf{x}_1$ and $\mathbf{x}_2$ both belong to $N^\perp$, then $\mathbf{x}^*(\mathbf{x}_1)\mathbf{x}_2 - \mathbf{x}^*(\mathbf{x}_2)\mathbf{x}_1$ belongs to N, showing that $N^\perp$ is one-dimensional.

Thus, any $\mathbf{x}$ in X can be written in the form $\alpha\mathbf{x}_0 + \mathbf{z}$, with $\mathbf{z}$ in N. Then, setting $\beta = \mathbf{x}^*(\mathbf{x}_0)$

$$\mathbf{x}^*(\mathbf{x}) = \alpha\beta = \langle\mathbf{x},\,\overline{\beta}\mathbf{x}_0\rangle.$$

which represents the linear functional as an inner product. If $\mathbf{x}^*$ were represented as inner products with two vectors $\mathbf{x}_1$ and $\mathbf{x}_2$, then for all $\mathbf{x}$, $\langle\mathbf{x},\,\mathbf{x}_1 - \mathbf{x}_2\rangle = 0$, the choice $\mathbf{x} = \mathbf{x}_1 - \mathbf{x}_2$ shows that $\mathbf{x}_1 = \mathbf{x}_2$. Any such inner product gives a bounded linear functional by the Cauchy-Schwarz inequality.   Q.E.D.

Corollary 3 *Let N be a closed subspace of a Hilbert space X and suppose that* $\mathbf{x}_0$ *does not belong to N. Then there is a continuous linear functional* $\mathbf{x}^*$, *defined on all of X, with* $\mathbf{x}^*(\mathbf{z}) = 0$ *for all* $\mathbf{z}$ *in N, and* $\mathbf{x}^*(\mathbf{x}_0) \ne 0$.

Proof It has been shown that there is a $\mathbf{z}_0$ in N with the properties that $\mathbf{w} = \mathbf{x}_0 - \mathbf{z}_0$ has $\|\mathbf{w}\| = d(\mathbf{x}_0,\,N)$ and that $\mathbf{w}$ is perpendicular to N. Defining $\mathbf{x}^*(\mathbf{x}) = \langle\mathbf{x},\,\mathbf{w}\rangle$ gives a bounded linear functional which is zero on N and which has $\mathbf{x}^*(\mathbf{x}_0) = \langle\mathbf{w}+\mathbf{z}_0,\,\mathbf{w}\rangle = \|\mathbf{w}\|^2 \ne 0$.   Q.E.D.

In computing solutions to physical problems, an approach often used is to show that the solution desired can be approximated to arbitrary accuracy by a class of functions which are easy to use in computations. The way this is often demonstrated to be the case, and which will be done in later chapters, is to consider the function you want to approximate, x_0, and the subspace N of functions you want to use to approximate x_0, and show that the approximation is possible by assuming that it is not, which is to say that x_0 does not belong to the closure of N. Then apply the Corollary above to obtain a continuous linear functional x^* which is zero on N and has $x^*(x_0) \neq 0$, showing that this leads to a contradiction. With this in mind, the goal of the next section is to establish Corollary 3 for a general Banach space.

3.7 The Hahn-Banach Theorem

Suppose that X is a Hilbert space, M a closed subspace of X, and x^*, is a continuous linear functional defined on M. By the Riesz Representation Theorem applied to the Hilbert space M, there is x_0 in M with $x^*(x) = \langle x, x_0 \rangle$ for all x in M. It is a simple matter to extend this linear functional x^* to a linear functional F defined on all of X by $F(x) = \langle x, x_0 \rangle$ for all x in X. And this extension has the same norm as the original functional: $\|F\| = \|x^*\| = \|x_0\|$. That this extension property is true for an arbitrary normed linear space is the content of the Hahn-Banach Theorem below.

An interesting feature of the proof is that it is first proved for Banach spaces over the real numbers, which allows the use of the order relation on the real numbers, and then extended to Banach spaces over the complex numbers.

Theorem 17 *The Hahn-Banach Theorem. Let M be a subspace of a normed linear space X over the real scalars and* x^* *a continuous linear functional defined on M. There is a continuous linear functional F, defined on all of X, which agrees with* x^* *on M, $F(x) = x^*(x)$ for all x in M, with the norm $\|F\|$ (as a operator on X) the same as the norm of* x^* *(on M).*

Proof Note that if M were a subspace which is not closed, any continuous linear functional could be extended to the subspace which is the closure of M without increasing its norm just by taking limits, so M may be taken to be closed.

Let x_0 be a point of X which is not in M. We first extend x^* by one dimension, to the subspace spanned by x_0 and M, i.e., to

$$Y = \text{span}(x_0, M) = \{\alpha x_0 + x : \ \alpha \text{ real and } x \text{ in } M\}.$$

On Y, define $F(\alpha x_0 + x) = \alpha F(x_0) + x^*(x)$. It is easy to see that F so defined is a linear functional on Y, and it remains to choose $F(x_0)$ so that $\|F\| = \|x^*\|$. Which is to say that we want

$$\|\alpha F(x_0) + x^*(x)\| \leq \|x^*\| \|\alpha x_0 + x\|$$

for all x in M, which does hold for $\alpha = 0$.

Since for $\alpha \neq 0$

$$\alpha F(\mathbf{x}_0) + \mathbf{x}^*(\mathbf{x}) = \alpha \left[F(\mathbf{x}_0) + \mathbf{x}^*(\mathbf{x}/\alpha) \right] \qquad (3.14)$$

it is enough to show that Eq. (3.14) holds for $\alpha = \pm 1$, which gives the two equations

$$F(\mathbf{x}_0) + \mathbf{x}^*(\mathbf{x}) \le \|\mathbf{x}^*\| \, \|\mathbf{x}_0 + \mathbf{x}\|$$

and

$$-F(\mathbf{x}_0) + \mathbf{x}^*(\mathbf{x}) \le \|\mathbf{x}^*\| \, \| - \mathbf{x}_0 + \mathbf{x}\|$$

Thus, it will be possible to find a suitable value for $F(\mathbf{x}_0)$ if

$$\|\mathbf{x}^*\| \, \|\mathbf{x}_0 + \mathbf{x}\| - \mathbf{x}^*(\mathbf{x}) \ge \mathbf{x}^*(\mathbf{x}) - \|\mathbf{x}^*\| \, \| - \mathbf{x}_0 + \mathbf{x}\|$$

that is if

$$2\mathbf{x}^*(\mathbf{x}) \le \|\mathbf{x}^*\| \, (\|\mathbf{x}_0 + \mathbf{x}\| + \| - \mathbf{x}_0 + \mathbf{x}\|)$$

for $\mathbf{x}$ in the subspace M, which is true as $\|\mathbf{x}_0 + \mathbf{x}\| + \| - \mathbf{x}_0 + \mathbf{x}\| \ge 2\|\mathbf{x}\|$.

At this point it has been shown that a bounded linear functional defined on a (closed) subspace M in a real Banach space X can be extended with preservation of norm to a subspace one dimension larger. To complete the proof requires the application of a fundamental axiom of mathematical logic, Zorn's Lemma. This lemma applies to situations where there is a partial order $\le$ on a set of elements $\mathcal{A}$. This order relation, $a \le b$, for a, b, and c in $\mathcal{A}$, has the properties: (1) (Reflexive) For all a, $a \le a$, (2) (Anti-symmetric) If $a \le b$ and $b \le a$, then $a = b$ (where the equal sign means that a and b are the same element in $\mathcal{A}$), (3) (Transitive) If $a \le b$ and $b \le c$, then $a \le c$.

A good example to have in mind is a set $\mathcal{A}$, say the set R^2, with the elements of $\mathcal{A}$ being all the subsets of $\mathcal{A}$, and the relation $a \le b$ denoting that the subset a is contained in the subset b. An important feature to note is that this order is indeed a *partial* order: given two subsets a and b, it is possible that they are not related via $\le$, since it might be that neither set is a subset of the other.

A set C of elements contained in a partially ordered set $\mathcal{A}$ is called a chain if given any two elements a and b in C either $a \le b$ or $b \le a$. An element c (in $\mathcal{A}$) is an upper bound for the chain C if $a \le c$ for all a in C. Finally, m is a maximal element for $\mathcal{A}$ if the only way that an element a in $\mathcal{A}$ can satisfy $m \le a$ is if $a = m$, that is to say that m is the "biggest element" among all the element to which m can be compared. Note that there may be more than one maximal element or none.

With these definitions in mind, Zorn's Lemma states that if $\mathcal{A}$ is a partially ordered set in which each chain has an upper bound, then there is at least one maximal element for $\mathcal{A}$.

To apply Zorn's Lemma in this proof, let the partially ordered space consist of all norm preserving extensions of $\mathbf{x}^*$ to a subspace N containing M, that is each element of $\mathcal{A}$ will consist of a pair (N, F), where N is a subspace containing M

and F is a bounded linear functional on N which is a norm preserving extension of the linear functional $\mathbf{x}^*$ (given on M); i.e., $F(x) = \mathbf{x}^*(x)$ for x in M and the norm of F on N is the same as the norm of $\mathbf{x}^*$ on M. Define an order on this space by $(N_1, F_1) \le (N_2, F_2)$ to mean that N is contained in N_1 which in turn is contained in N_2, and F_1 and F_2 are, both norm preserving extensions of $\mathbf{x}^*$ to, respectively, N_1 and N_2, and moreover $F_1(x) = F_2(x)$ for x in N_1.

Let C be a chain in this space consisting of pairs (N, F) as indicated. Let $\tilde{N}$ be the union of all the sets in the chain, and define a linear functional $\tilde{F}$ on $\tilde{N}$ as follows: any $\mathbf{x}$ in $\tilde{N}$ belongs to some N in the chain. Use this pair (N, F) in the chain to define $\tilde{F}(x) = F(x)$. Because C is a chain, the definition of the order relation shows that the same value for $\tilde{F}(x)$ will be obtained no matter which subspace N in the chain contains $\mathbf{x}$. It is clear that $(\tilde{N}, \tilde{F})$ is an upper bound for the chain.

Applying Zorn's Lemma shows that the partially ordered set $\mathcal{A}$ has a maximal element (N_0, F_0). Then the subspace N_0 is closed, for if not F_0 could be extended to a larger subspace, the closure of N_0, by taking limits and the resulting pair would be larger than the maximal (N_0, F_0). Finally, N_0 must be all of the normed linear space X, since if it were a proper subspace, it has been shown that the linear functional F_0, itself being a norm preserving extension of $\mathbf{x}^*$, can be extended by one dimension to a larger subspace, and that would contradict the maximality of (N_0, F_0).   Q.E.D.

In order to prove the Hahn-Banach Theorem for normed linear spaces with scalars the complex numbers a lemma is needed. By the phrase "a real linear functional on a real vector space V" is meant a map f of V into the real numbers with $f(u + v) = f(u) + f(v)$, for u and v in V, and (the property of interest here) $f(au) = af(u)$ for u in V and any real number a. Also, if V is a complex vector space, a real linear functional satisfies exactly the same requirements; but in this case for a complex number $a + ib$, a and b real, we do not know that $f((a+ib)u) = (a+ib)f(u)$, but we do know that $f(au) = af(u)$ for real numbers a. On the other hand, a complex linear functional F on a complex vector space is additive and has $F((a + ib)u) = (a + ib)F(u)$ for all u in V and all complex scalars $a + ib$. With this terminology:

Lemma 11 *Let V be a vector space over the complex numbers. For F a complex linear functional on V, consider the real and imaginary parts of the complex values of F. Then for all u in V with $F(u) = ReF(u) + iImF(u)$, ReF and ImF are real linear functionals on V and $ReF(iu) = -Im(u)$.*

Conversely, if G(u) is a real linear functional on V, then

$$F(u) = G(u) - iG(iu)$$

is a complex linear functional on V.

Proof First, it is clear that ReF and ImF are real linear functionals, and

$$ReF(iu) = Re(iF(u)) = -ImF(u).$$

Second

$$F((a+ib)u) = G(au) + G(ibu) - i(G(iau) + G(-bu)) =$$

$$aG(u) + ibG(u) - iaG(iu) - ibG(u) = (a+ib)(G(u) - iG(iu))$$

Q.E.D.

Theorem 18 *Hahn-Banach Theorem, Complex Scalars. Let* $\mathbf{x}^*$ *be a (complex) continuous linear functional defined on a closed subspace M of a normed linear space X with scalar field the complex numbers. Then there is a continuous linear functional F defined on all of X with* $F(\mathbf{x}) = \mathbf{x}^*(\mathbf{x})$ *for all* $\mathbf{x}$ *in M and* $\|F\| = \|\mathbf{x}^*\|$.

Proof Since X is also a real normed linear space, applying Theorem 17 shows there is a real linear functional G defined on X which equals $\mathbf{x}^*$ on M and has $\|G\| = \|\mathbf{x}^*\|$. By the above lemma, defining $F(u) = G(u) - iG(iu)$ gives a complex linear functional which agrees with $\mathbf{x}^*$ on M. Further, if $F(\mathbf{x}) = re^{i\theta}$, r real, then $|F(\mathbf{x})| = |r| = |e^{-i\theta}F(\mathbf{x})| = |F(e^{-i\theta}\mathbf{x})| = |G(e^{-i\theta}\mathbf{x})| \le \|\mathbf{x}^*\|\|e^{-i\theta}\mathbf{x}\| = \|\mathbf{x}^*\|\|\mathbf{x}\|$.   Q.E.D.

Corollary 4 *Let M be a closed subspace of a normed linear space X (over either the real or complex scalars) and* $\mathbf{x}_0$ *a point in X which is not in M. There is a continuous linear functional* $\mathbf{x}^*$ *defined on X with* $\mathbf{x}^*(\mathbf{x}) = 0$ *for all* $\mathbf{x}$ *in M,* $\mathbf{x}^*(\mathbf{x}_0) = 1$, *and* $\|\mathbf{x}^*\| = \frac{1}{d}$, *where* $d = d(\mathbf{x}_0, M)$.

Proof Consider the subspace spanned by $\mathbf{x}_0$ and vectors from M, that is all vectors of the form $\alpha\mathbf{x}_0 + \mathbf{z}$ for $\mathbf{z}$ in M and α a scalar. On this subspace, define a functional F by $F(\alpha\mathbf{x}_0 + \mathbf{z}) = \alpha$. Using the fact that M is a subspace it is easy to show that F is a linear functional. From $\|\alpha\mathbf{x}_0 + \mathbf{z}\| = |\alpha|\|\mathbf{x}_0 + \mathbf{z}/\alpha\| \ge d|\alpha|$, it follows that $\|F\| \le \frac{1}{d}$ on the subspace spanned by $\mathbf{x}_0$ and M. Choosing $\mathbf{z}_n$ in M with $\|\mathbf{x}_0 - \mathbf{z}_n\| \to d$, $1 = |F(\mathbf{x}_0 - \mathbf{z}_n)| \le \|F\|\|\mathbf{x}_0 - \mathbf{z}_n\| \to \|F\|d$, shows that the norm on F on the span of $\mathbf{x}_0$ and M is precisely $\frac{1}{d}$. Now apply the Hahn-Banach Theorem.   Q.E.D.

3.8 Exercises

1. Show that l^1 is a Banach space.
2. Consider the subspace of l^1 consisting of all the sequences which have only a finite number of non-zero terms. This is a normed linear space. Show that it is not a Banach space.
3. Define a linear functional on C[0,1] by $x^*(f) = f(0)$. Show that x^* is a bounded linear functional and that there is no Riemann integrable function h with $x^*(h) = \int_0^1 f(t)h(t)dt$.
4. For X and Y normed linear spaces, if the space of bounded linear operators from X to Y, B(X, Y) is complete, then Y is complete.
5. Give an example of a bounded sequence of functions in C[0,1] which do not have a convergent subsequence.

6. Let $\mathbf{x}_0$ be an element of a normed linear space X. Show that $\|\mathbf{x}_0\| = \sup\{\mathbf{x}^*(\mathbf{x}_0) :$ $\|\mathbf{x}^*\| = 1\}$, the sup being taken over all norm one linear functionals in X^*.

7. Let c_0 denote the subspace of l^∞ consisting of sequences which converge to zero. Prove that c_0 is a Banach space whose conjugate space is l^1.

Chapter 4
Power Series

Abstract Power series in the complex variable z are defined. Convergence is discussed, and the formula for the radius of convergence is derived. Example given include e^z, $cos(z)$, and $sin(z)$. It is shown that these series can be differentiated term-by-term, a consequence of which is the Cauchy-Riemann equations which imply that in its circle of convergence a power series $f(z)$ has harmonic real and imaginary parts, a fact based on the complex variable boundary element method. A proof is given of the Weierstrass Approximation Theorem that every continuous function defined on a finite closed interval can be there uniformly approximated by polynomials

Keywords Power Series · Absolute convergence · Radius of convergence · Ratio test · Convergence of the derivative series · The exponential function · Weierstrass Approximation Theorem.

A power series is a series in the complex variable z

$$\sum_0^\infty c_n z^n$$

with the coefficients c_n complex numbers.

Definition 22 For a sequence of non-negative real numbers $\{a_n\}$, the *limit supremum* of the sequence, written *limsup* a_n, or $\overline{\lim} \, a_n$, is defined by

$$\lim_{n\to\infty} \sup_{m\geq n} a_m.$$

This limit exists as an extended real number. It may be $+\infty$ or, if not, it is a non-negative real number. To see this, the sequence $A_n = \sup_{m\geq n}\{a_m\}$ is non-increasing. If it is not $+\infty$ for all n, then, if finite for n_0, it is finite for all larger n. It is then easy to see that A_n, being bounded below (by zero), converges, in fact to the infimum of the set $\{A_n : n \geq n_0\}$

T. Hromadka and R. Whitley, *Foundations of the Complex Variable Boundary Element Method*, SpringerBriefs in Applied Sciences and Technology, DOI: 10.1007/978-3-319-05954-9_4, © The Author(s) 2014

Lemma 12 *A series $\sum b_n$ of complex numbers converges if $\sum |b_n|$ converges; absolute convergence implies convergence.*

Proof For $m_1 < m_2$,

$$\left| \sum_0^{m_2} b_n - \sum_0^{m_1} b_n \right| = \left| \sum_{m_1+1}^{m_2} b_n \right| \leq \sum_{m_1}^{m_2} |b_n| = \sum_0^{m_2} |b_n| - \sum_0^{m_1} |b_n|$$

and the right-hand side converges to zero because the partial sums of the convergent series $\sum |b_n|$ are Cauchy. Note that one consequence of the partial sums being Cauchy is that the sum of the first n terms minus the sum of the first $n-1$ terms, which is to say the nth term b_n, must go to zero.   Q.E.D.

Theorem 19 *For the power series $\sum_0^\infty c_n z^n$, the radius of convergence R is defined to be the reciprocal of $L = \limsup |c_n|^{\frac{1}{n}}$.*

If $L = \infty$, R is taken to be zero, and the series only converges for the single point $z = 0$.

If $L = 0$, R is taken to be ∞. In this case, the series converges for all z.

If L is a positive real number, $R = 1/L$ and the series converges for $|z| < R$ and does not converge for $|z| > R$.

Proof Suppose that $R = 0$. Let a nonzero value of z be given. Then $|c_n|^{\frac{1}{n}} > \frac{1}{|z|}$ for infinitely many values of n. Thus, the nth term of the series does not converge to zero and the series therefore diverges.

Suppose that $R = \infty$. For any nonzero z and $\rho < 1$, there is an integer N with $|c_n|^{\frac{1}{n}} \leq \frac{\rho}{|z|}$ for $n \geq N$. Then $\sum_N^\infty |c_n z^n| \leq \sum_N^\infty \rho^n$, which is finite being the tail end of a convergent geometric series.

Suppose that $0 < R < \infty$. If $0 < |z| < R$, then $\frac{1}{|z|} > L$, and there is a real number $\rho < 1$ satisfying $\frac{1}{|z|} > \frac{\rho}{|z|} > L$. Then $|c_n|^{\frac{1}{n}} < \frac{\rho}{|z|}$ for large n, whence $|c_n||z|^n < \rho^n$ for large n, and so by comparison with the convergent geometric series $\sum \rho^n$ the power series in z converges. On the other hand, if $|z| > R$, then $|c_n|^{\frac{1}{n}} > \frac{1}{|z|}$ for infinitely many n, showing that the nth term $c_n z^n$ does not tend to zero and therefore the power series does not converge.   Q.E.D.

This result shows that $\sum c_n z^n$ and $\sum |c_n| z^n$ have the same radius of convergence. Note that the theorem provides no information about the convergence or divergence of the series for those z with $|z| = R$.

Examples

(1) $\sum \frac{z^n}{n^2}$. The radius of convergence $R = 1$ and the series converges for all z with $|z| = 1$.

(2) $\sum n z^n$, has $R = 1$ and diverges for all $|z| = 1$ since for such z the nth term does not go to zero.

(3) The series $\sum \frac{z^n}{n}$ diverges at $z = 1$. It converges for all z with $|z| = 1$ and $z \neq 1$, but the proof is tricky.

(4) One of the most useful mathematical functions is e^z, with power series $e^z = \sum \frac{z^n}{n!}$. The radius of convergence R can be computed from the formula $L = \lim[\frac{1}{n!}]^{1/n}$, but see the easier computation after the next lemma.

Lemma 13 *Ratio Test*

For the series $\sum_1^\infty a_n$ of nonzero terms:

If $\lim \frac{|a_{n+1}|}{|a_n|}$ *exists and is less than one, the series converges.*

If $\lim \frac{|a_{n+1}|}{|a_n|}$ *exists and is greater than one, the series diverges.*

Proof Choose $\rho < 1$ with $\lim \frac{|a_{n+1}|}{|a_n|} < \rho$. There is an N with $\frac{|a_{n+1}|}{|a_n|} < \rho$ for $n \geq N$. Then

$$a_{N+m} \leq a_{N+m-1} \rho \cdots \leq a_N \rho^m.$$

By comparison with the geometric series, $\sum a_n$ converges.

In the second case, choose $1 < \rho < \lim \frac{|a_{n+1}|}{|a_n|}$. Argue as above with the inequality reversed to see that

$$a_{N+m} \geq \rho a_{m-1+N} \rho \cdots \geq a_N \rho^m$$

which shows that a_n does not tend to zero and so the series cannot converge. Q.E.D.

Apply the ratio test to the series for e^z. For the series $\sum_0^\infty \frac{|z|^n}{n!}$, apply the ratio test to the ratio $\frac{|z|^{n+1}}{(n+1)!} / \frac{|z|^n}{n!} = \frac{|z|}{n+1}$ which tends to zero as n tends to infinity, showing that the series converges (absolutely) for all z.

Lemma 14 *If the power series $\sum_0^\infty c_n z^n$ has radius of convergence $R > 0$, the series convergences uniformly and absolutely for $|z| \leq r < R$.*

Proof First suppose that $R = \infty$, so that $\limsup |c_n|^{1/n} = 0$. Given r and $\epsilon_0 > 0$ there is an N with $|c_n|^{1/n} < \frac{1}{r+\epsilon_0}$, for $n \geq N$. Then for $|z| \leq r$

$$\sum_N^\infty |c_n||z|^n \leq \sum_N^\infty \left[\frac{r}{r+\epsilon_0}\right]^n$$

the right-hand side being a convergent geometric series. It follows that the power series convergences absolutely and uniformly for $|z| \leq r$ since the tail end of the series can be made arbitrarily small for all $|z| \leq r$.

If R is finite, $|c_n|^{1/n} \leq \frac{1}{R}$ and $\sum |c_n||z|^n \leq \sum (r/R)^n$.   Q.E.D.

Since the partial sums of a power series are polynomials, and so are continuous, and converge uniformly to $f(z)$ for $|z| \leq r < R$, $f(z)$ is continuous; much more is true.

Theorem 20 *Let $f(z) = \sum_0^\infty c_n z^n$ converge for $|z| < R$. Then $f(z)$ is differentiable for $|z| < R$ with derivative $f'(z)$ given by*

$$f'(z) = \sum_1^\infty n c_n z^{n-1},$$

and the series for $f'(z)$ also converges for $|z| < R$.

Proof Note that $\limsup [n|c_n|]^{1/n} = 1/R$, since $\lim n^{1/n} \to 1$. (To see this, take the logarithm and use L'Hopital's rule to compute

$$\lim_{x \to \infty} \frac{\log(x)}{x} = \lim_{x \to \infty} \frac{1}{x} = 0.$$

showing that the differentiated series has the same radius of convergence as the original series.

Consider the difference quotient of f, with step h, which is a complex number, minus the differentiated series:

$$\frac{f(z+h) - f(z)}{h} - \sum_1^\infty n c_n z^{n-1} = \sum_1^\infty c_n \left[\frac{(z+h)^n - z^n}{h} - n z^{n-1} \right].$$

Look at the expression in bracket which multiplies c_n in the above

$$\left[\frac{(z+h)^n - z^n}{h} - n z^{n-1} \right] = \frac{h}{z^2} \sum_{j=2}^n \binom{n}{j} h^{j-2} z^{n-(j-2)}. \qquad (4.1)$$

Now $\binom{n}{j} = \frac{n(n-1)\dots(n-j+1)}{j!} \le n^2 \binom{n}{j-2}$, for $j \ge 2$.
so the expression above is bounded in absolute value by

$$\frac{|h|}{|z|^2} n^2 \sum_0^n \binom{n}{j} |h|^j |z|^n = \frac{n^2 |h|}{|z|^2} (|h| + |z|)^n \le n^2 \frac{|h|}{|z|^2} (R - \delta)^n,$$

for small positive δ. This gives

$$\left| \frac{f(z+h) - f(z)}{h} - \sum_1^\infty n c_n z^{n-1} \right| \le \frac{|h|}{|z|^2} \sum_0^\infty n^2 |c_n| (|z| + |h|)^n.$$

Since the radius of convergence of $\sum n^2 c_n z^n$ and $\sum c_n z^n$ are the same, for z fixed and not zero, difference quotient (4.1) is bounded by a constant A times $|h|$.

If R is infinite, taking $|h| \le 1$, gives a bound

$$|\frac{f(z+h) - f(z)}{h} - \sum_1^\infty n c_n z^{n-1}| \le \frac{|h|}{|z|^2} \sum_0^\infty n^2 |c_n|(|z| + 1)^n$$

while if R is finite, for small δ, with $|z| \le R - 2\delta$, and $|h| \le \delta$,

$$|\frac{f(z+h) - f(z)}{h} - \sum_1^\infty n c_n z^{n-1}| \le \frac{|h|}{|z|^2} \sum_0^\infty n^2 |c_n|(R - \delta)^n.$$

When $h \to 0$, in either case the right-hand side tends to zero.   Q.E.D.

Example 11 The function $e^z = \sum_0^\infty \frac{z^n}{n!}$, converging for all z, has $f'(z) = \sum_1^\infty \frac{n z^{n-1}}{n!} = \sum_0^\infty \frac{z^n}{n!} = e^z$.

Corollary 5 *A function defined by a power series converging for $|z| < R$ is infinitely differentiable, the kth derivative being obtained by differentiating the series term by term k times, the series obtained also converging for $|z| < R$.*

Proof Use induction and Theorem 20. Applying the Corollary and evaluating the kth derivative at zero gives $f^{(k)}(0) = k! c_k$ and which gives the formula for c_k:

$$c_k = \frac{f^{(k)}(0)}{k!}. \tag{4.2}$$

Q.E.D.

Example 12 If a power series satisfies $f^{(k)}(z) = f(z)$ for all $k = 1, 2, \ldots$, then $c_k = k! f^{(k)}(0) = k! f(0)$, and $f(z) = c e^z$, $c = f(0)$. In fact if all the derivatives of $f(z)$ are equal for z equals zero, then $f(z) = c e^z$.

Example 13 Start with the series for $\sin(x)$ and $\cos(x)$, x real, and substitute complex z for x. Or use formula 4.2 and assume that sine and cosine have a power series expansion:

$$\sin(z) = z - \frac{z^3}{3!} + \frac{z^5}{5!} - \frac{z^7}{7!} + \cdots$$

and

$$\cos(z) = 1 - \frac{z^2}{2!} + \frac{z^4}{4!} - \cdots$$

with the series converging for all z by comparing the series for $\sin(z)$ and $\cos(z)$ with the series for e^z. These series extend the definitions of $\cos(x)$ and $\sin(x)$ to complex values of the argument. Differentiating the series shows that the derivative of $\sin(z)$ is $\cos(z)$ and that the derivative of $\cos(z)$ is $-\sin(z)$, for complex z.

Substituting $i\theta$, for θ real, into the series for $e^{i\theta}$ gives the famous relation:

$$e^{i\theta} = \cos(\theta) + i\,\sin(\theta), \tag{4.3}$$

and the consequence $e^{i\pi} = -1$.

Example 14 For the convergent power series $f(x) = \sum a_n z^n$, for $|z| < R_1$ and $g(z) = \sum b_n z^n$, for $|z| < R_2$ the product is again a power series $f(z)g(z) = \sum c_n z^n$, for $|z| < \min(R_1, R_2)$. Multiplying the two series and collecting the coefficient of z^n gives $c_n = \sum_0^n a_j b_{n-j}$.

For $f(z) = e^{\alpha z}$ and $g(z) = e^{\beta z}$,

$$c_n = \sum_0^n \frac{\alpha^j \beta^{n-j}}{j!(n-j)!} = \frac{1}{n!} \sum_0^N \binom{n}{j} \alpha^j \beta^{n-j} = \frac{(\alpha+\beta)^n}{n!}.$$

Which shows that $e^{\alpha z} e^{\beta z} = e^{(\alpha+\beta)z}$, and setting $z = 1$,

$$e^\alpha e^\beta = e^{\alpha+\beta}.$$

Using the above equation and Eq. 4.3 for α and β real and comparing real and imaginary parts gives the addition formulas

$$\cos(\alpha + \beta) = \cos(\alpha)\cos(\beta) - \sin(\alpha)\sin(\beta)$$

$$\sin(\alpha + \beta) = \cos(\alpha)\sin(\beta) + \sin(\alpha)\cos(\beta).$$

Theorem 21 *Weierstrass Approximation Theorem.*

Proof Since any finite closed interval can be mapped into another finite closed interval by a linear map, and substituting a linear map in the argument of a polynomial gives another polynomial, we can take the interval to be any convenient interval; we choose $[-\pi, \pi]$. Given the continuous function f on this interval, set

$$g(x) = f(x) + \frac{x+\pi}{2\pi}(f(-\pi) - f(\pi)).$$

This continuous function g has $g(-\pi) = g(\pi)$, and so it can be extended to be periodic with period 2π over the entire real line R. And, if there is a polynomial which uniformly approximates g to within ϵ, it is easy to add a linear term to this polynomial and so uniformly approximate f to within ϵ.

By Theorem 4, letting $S_j(x)$ denote the jth partial sum of the Fourier series for $g(x)$, the average

$$A_N(x) = \frac{1}{N+1}[S_0(x) + S_1(x) + \cdots + S_N(x)]$$

will for large enough N approximate g uniformly on $[-\pi, \pi]$ to within a given ϵ. Write out A_N as

$$A_N(x) = a_0 + \sum_1^N a_j \cos(jx) + \sum_1^N b_j \sin(jx).$$

and set $M = \sum_1^N |a_j| + \sum_1^N |b_j|$. On any finite interval, in fact over any finite circle centered at zero in the plane, the power series for $\cos(x)$ and $\sin(x)$ converge uniformly. On the interval $[-N\pi, N\pi]$, the functions $\cos(x)$ and $\sin(x)$ can be approximated to arbitrary accuracy by partial sums $P(x)$ and $Q(x)$ of their power series, which are polynomials: $|\cos(x) - P(x)| < \frac{\epsilon}{M}$ and $|\sin(x) - Q(x)| < \frac{\epsilon}{M}$. Then

$$|a_0 + \sum_1^N a_j P(jx) + \sum_1^N b_j Q(jx) - A_N(x)| \le \epsilon.$$

and the sums in the above equation plus the constant a_0 constitute a polynomial which approximates $g(x)$ uniformly to within 2ϵ.   Q.E.D.

Theorem 22 *If the power series $f(z) = \sum c_n z^n$ converges with $R > 0$, and there is an infinite sequence of points $z_n \to 0$ with $f(z_n) = 0$ for all n, then the power series is identically zero.*

Proof First $c_0 = f(0) = \lim f(z_n) = 0$. So the series has the form $f(z) = c_1 z + c_2 z^2 + \cdots$ which can be written $\frac{f(z)}{z} = c_1 + c_2 z + \cdots$, and as this series is continuous at zero the same argument as before shows that $c_1 = 0$, and $\frac{f(z)}{z^2} = c_2 + c_3 z + \cdots$. Proceeding inductively, all the coefficients c_n are zero.   Q.E.D.

Example 15 It is known from calculus that $\sin^2(x) + \cos^2(x) = 1$ for real x. The powers series for $\sin^2(z)$ and $\cos^2(z)$ are seen to converge since they are obtained by squaring the partial sums of their respective power series, which converge absolutely for $|z| \le r < \infty$, and then taking a limit. The power series $\sin^2(z) + \cos^2(z) - 1$ is zero for all real z, and so for any real sequence of points converging to zero, hence is identically zero for all z, i.e., $\sin^2(z) + \cos^2(z) = 1$. The same argument can be used to extend various identities known for real x to the complex plane; an example, discussed, above being the addition formulas for $\sin(z + w)$ and $\cos(z + w)$.

The results for a power series in powers of z can be easily extended to a series $g(z) = \sum_0^\infty c_n(z - z_0)^n$. For example, $c_n = \frac{g^{(n)}(z_0)}{n!}$. Under the change of variable $w = z - z_0$, it is clear that the series converges in the disk $|z - z_0| < R$ with $\frac{1}{R} = \limsup |c_n|^{1/n}$.

4.1 Cauchy-Riemann Equations

If f is a function with a power series and z_0 is in the circle of convergence, we know that $f(z)$ is differentiable at z_0. The derivative at z_0 is given by the limit of the difference quotient $\frac{f(z_0+h)-f(z_0)}{h}$ as h tends to zero through complex values. Taking

h to have complex values has important consequences. Write f in terms of its real part $U = Re(f)$ and imaginary part $= Im(f) = V$. Writing $z = x + iy$, in terms of the real and imaginary parts of z, the functions U and V can be regarded as functions of the two real variables x and y:

$$f(z) = f(x + iy) = U(x, y) + iV(x, y).$$

Since f is differentiable at z_0, the difference quotient tends to $f'(z_0)$ as $h \to 0$. If h is taken to have only real values as it approaches $z_0 = x_0 + iy_0$:

$$\lim_{h \to 0} \frac{U(x_0 + h, y_0) - U(x_0, y_0)}{h} + i \frac{V(x_0 + h, y_0) - V(x_0, y_0)}{h}$$

The first term converges to the partial derivative of U with respect to x, at (x_0, y_0), $U_x(x_0, y_0)$, while the second term converges to the partial derivative of V with respect to x at (x_0, y_0):

$$f'(z_0) = U_x(x_0, y_0) + iV_x(x_0, y_0).$$

If instead we let h tend to zero through pure imaginary values, ih, with h real, a similar calculation shows that in terms of the partial derivatives with respect to y, U_y and V_y,

$$f'(z_0) = \lim_{h \to 0} \frac{U(x_0, y_0 + h) - U(x_0, y_0)}{ih} + i \frac{V(x_0, y_0 + h) - V(x_0, y_0)}{ih}$$

and

$$f'(x_0, y_0) = V_y(x_0, y_0) - iU_y(x_0, y_0).$$

Writing the two forms of $f'(z_0)$ $f' = U_x + iV_x = V_y - iU_y$, in terms of the real and imaginary parts

$$U_x = V_y \text{ and } U_y = -V_x, \tag{4.4}$$

which are the Cauchy-Riemann equations.

The functions U_x, U_y, V_x, and V_y are real or imaginary parts of the power series $f'(z)$ and so are continuous and, as above, have partial derivatives. Since $f'(z) = U_x(x, y) + iV_x(x, y)$, $U_x = Re(f')$ and $V_y = Im(f')$. Applying the Cauchy-Riemann equations for $f'(z)$ gives $U_{xx} = V_{xy}$ (As an example of the notation, taking the partial of V_y with respect to x, $(V_y)_x$ by V_{yx}). But also $f'(z) = V_y - iU_x$, and therefore $V_{yx} = -U_{xx}$. By a theorem in calculus, $V_{xy} = V_{yx}$, and

$$\Delta U = U_{xx} + U_{yy} = 0$$

showing that U has continuous partial derivatives and satisfies Laplace's equation in the domain where f is differentiable, so is an harmonic function. By a similar computation, or since $V = \text{Re}(-if)$, V is also harmonic. The real and imaginary parts of a function with a power series will be used in the next two chapters to approximate solutions to the Dirichlet problem for the heat equation.

If a function $f(z)$ has a derivative at all points in open set, then it has derivatives of all orders and has a convergent power series about the points in this set. Such functions are analytic and this remarkable result, that the existence of one derivative implies the existence of all the derivatives and implies the existence of a convergent power series, is one of the many strong results in the study of analytic functions. See, for example, [1].

4.2 Exercises

1. Consider the series $\sum a_n$ of positive terms.
   (a) Show that if $limsup\{\frac{a_{n+1}}{a_n}\} < 1$ the series converges. (b) For a sequence of non-negative numbers $\{b_n\}$, define $liminf\, b_n$ (c) Show that if $liminf\{\frac{a_{n+1}}{a_n}\} > 1$ the series diverges.
2. Let θ belong to $[\pi, \pi]$, use

$$e^{in\theta} = \cos(n\theta) + i\,\sin(n\theta)$$

to obtain the expressions:

$$\cos(n\theta) = \frac{e^{in\theta} + e^{-in\theta}}{2} \quad \text{and} \quad \sin(n\theta) = \frac{e^{in\theta} - e^{-in\theta}}{2}.$$

The Fourier series $A_0/2 + \sum A_n\cos(n\theta) + \sum B_n\sin(n\theta)$ can be written in the form $\sum_{-\infty}^{\infty} C_n e^{i\,n\theta}$ with complex coefficients C_n. This form is often used, see [2].
3. Prove that $\lim_{n\to\infty} n^{1/n} = 1$. Prove that $\lim_{n\to\infty}(1/n!)^{1/n} = 0$
4. What is the function $f(z)$ whose power series for $|z| < 1$ is $\sum n^2 z^n$?
5. Show that $\sin(x + iy) = \sin(x)\,\cosh(y) + i\,\cos(x)\,\sinh(y)$, using the hyperbolic functions $\cosh(y) = \frac{e^y + e^{-y}}{2}$ and $\sinh(y) = \frac{e^y - e^{-y}}{2}$.
6. Consider the function $h(x) = \frac{1}{1+x^2}$. Find its power series. If only real values for x are considered it is not clear why the series does not converge for $|x| > 1$. What is clear when $h(z)$ is regarded as a function of the complex variable z?

References

1. Bak J, Newman J (1982) Complex analysis. Springer, Berlin
2. Körner T (1988) Fourier analysis. Cambridge University Press, Cambridge

Chapter 5
The R^2 Dirichlet Problem

Abstract It is shown how an approximate solution to a Dirichlet problem-a real-valued continuous function g being given on the boundary of a bounded domain D without holes-can be constructed using any one function $f(z)$, not a polynomial, which has a convergent power series at some point. (And it is shown that if f is a polynomial, the result is false.) The proof uses the Hahn-Banach Theorem. If the function f is taken to have a power series convergent about zero, the solution, harmonic in D and approximating g on the boundary, is the real part of a finite sum of terms of the form $c_n(f(a_n(z + b_n))$. In applications the real parameters a_n, b_n, c_n, $n = 1, 2, \ldots, M$ are obtained by computer minimization of the error of the fit to the given function g on the boundary of the domain. The proof depends on the Walsh-Lebesgue Theorem, which is shown to be equivalent to the complex variable approximate boundary value result as stated in this chapter. Note that the approximation is to the boundary function g on the boundary of the domain, the approximating functions satisfy the Laplace equation exactly in D .

Keywords Hole in a domain · Walsh-Lebesgue Theorem · C(Γ) · Hahn-Banach Theorem.

The heat equation in the unit square in the plane, considered in Chap. 1, is the prototype of the problem discussed in this chapter. The domain will be a bounded open subset of R^2 with boundary Γ. The boundary condition, the temperature at points of the boundary, will be described by a continuous function g which maps the boundary of D, $\partial D = \Gamma$, into the real numbers $g : \Gamma \rightarrow R$. One further condition is that the domain have no holes. One way to make that precise is as follows:

Definition 23 An arc $\mathcal{A}$ in R^2 is a continuous image of an interval, i.e., there is a continuous function $\Lambda : [0, 1] \rightarrow R^2$, mapping $[0, 1]$ onto $\mathcal{A}$. An open set U in R^2 is arcwise-connected if given any two points a and b in U, there is an arc given by $\Lambda : [0, 1] \rightarrow U$, (so with image lying entirely in U), $\Lambda(0) = a$, and $\Lambda(1) = b$.

The domain D in R^2 is said to have no holes if its closure $K = D \cup \Gamma$ has complement $R^2 - K$ which is arcwise-connected.

T. Hromadka and R. Whitley, *Foundations of the Complex Variable Boundary Element Method*, SpringerBriefs in Applied Sciences and Technology, DOI: 10.1007/978-3-319-05954-9_5, © The Author(s) 2014

For example, the doughnut $D = \{z : 1 < |z| < 2\}$ does indeed have a hole, as both the circle $\{|z| < 1\}$ and the set $\{|z| > 2\}$ belong to the complement of the closure of K and points in one cannot be joined to points in the other by an arc lying in the complement.

The proof of the theorem of this chapter will use the following theorem, stated without proof.

Theorem 23 *[1, p. 173] Walsh-Lebesgue Theorem. Let K be a compact subset of R^2 with $R^2 - K$ arcwise-connected. Then for any continuous function g mapping the boundary of K into the real numbers, $g : \Gamma \to R$, there are polynomials in z with real part uniformly close to g on Γ, i.e., given any $\epsilon > 0$, there is a polynomial $p(z)$ with*

$$|Re(p(z)) - g(z)| < \epsilon \text{ for all } z \text{ on } \Gamma$$

Since the polynomial $p(z) = \sum_{j=0}^{N} c_n z^j$ is a (finite) power series, its real and imaginary parts satisfy Laplace's equation, and so the function $Re(p(z))$ of the Walsh-Lebesgue Theorem is harmonic, satisfying the heat equation in the domain, and has values on the boundary which are approximately equal to the boundary conditions given by g.

This important theoretical result does not generally give acceptable numerical results. As most Numerical Analysis textbooks emphasize, there are problems with polygonal approximation on an interval, for example with interpolating polynomials, and these problems are worse when trying to approximate on a curve in R^2, and worse yet in R^3. In contrast, the methods of this chapter will be generalized in the next chapter so as to be applicable to three-dimensional problems, and even to problems in R^N.

The theorem below gives a method, based on any one function f with a power series expansion which is not a polynomial, for approximating the solution to a Dirichlet problem. (An example after the theorem will show that the restriction that the function not be a polynomial is necessary).

Theorem 24 *Let f be have a power series convergent (for simplicity of notation about zero) for $|z| < \rho$ which is not a polynomial. An approximate solution to the Dirichlet problem for any bounded open domain without holes and any continuous function g mapping the boundary of the domain to the reals, can be constructed using sums of this one function in terms of the form $f(a(z+b))$. The only restrictions on the parameters a and b is that small values are admissible and that the argument $a(z+b)$ lie in a closed circle inside the circle of convergence for f. For example, consider D a bounded open set, say $|z| \le M$ for z in D, with closure K, with no holes (i.e., $R^2 - K$ arcwise-connected). Suppose that g is a continuous function mapping the boundary $\partial D = \Gamma$ into the real numbers R. If the power series for f converges for $|z| < \rho$—if convergent for all z, take ρ to be some positive real number—restricting the parameters a and b by*

$$0 \le b \le M \text{ and } 0 \le a \le \frac{\rho}{4M}, \tag{5.1}$$

will do, although other choices may be better for numerical work.

Then given $\epsilon > 0$, there is a finite sum,

$$h(z) = \sum_{0}^{N} c_n f(a_n(z + b_n)) \tag{5.2}$$

a_n, b_n *real numbers with all the a_n satisfying the condition on a in Eq. (5.1) and all the b_n satisfying the conditions on b in (5.1), and the c_n complex numbers, with*

$$|Re h(z) - g(z)| < \epsilon$$

for all z on Γ,

Proof The argument will take place in the Banach space $C(\Gamma)$ of continuous (complex-valued) functions on the compact metric space Γ with the supremum norm. Let $\mathcal{M}$ be the subspace of $C(\Gamma)$ spanned by all the functions of the form $f(a(z + b))$, a and b restricted as indicated in (5.1). By The Lebesgue-Walsh Theorem, given $\epsilon > 0$ there is a polynomial $p(z)$ with $|Re p(z) - g(z)| < \epsilon$. If $p(z)$ belongs to the closure of $\mathcal{M}$, then there would be a function $h(z)$ as in the statement of the theorem, with $|h(z) - p(z)| < \epsilon$, and the theorem would follow as then $|Re h(z) - g(z)| < 2\epsilon$.

Suppose instead that there is a polynomial $p(z)$ which does not belong to the closure of $\mathcal{M}$. By the Hahn-Banach Theorem there is a continuous linear functional $\mathbf{x}^*$ defined on $C(\Gamma)$ with

$$\mathbf{x}^*(\mathcal{M}) = 0, \text{ and } \mathbf{x}^*(p) \neq 0.$$

and since the value of $\mathbf{x}^*(p)$ is merely specified not to be zero, it can be supposed that $\|\mathbf{x}^*\| = 1$. The power series $f(w) = \sum \gamma_n w^n$ converges uniformly for all the $w = a(z + b)$ under consideration because such w have $|w| \leq \frac{\rho}{2}$ and power series converge uniformly on closed disks smaller than the disk of convergence. Hence, by the continuity of $\mathbf{x}^*$ (with respect to the sup norm of $C(\Gamma)$) and the fact that the terms $f(a(z + b))$ are in $\mathcal{M}$

$$0 = \mathbf{x}^*(f(a(z + b)) = \sum \gamma_n \mathbf{x}^*[(a(z + b))^n] = \sum \gamma_n a^n \mathbf{x}^*[(z + b)^n].$$

The right-hand side of the above is a power series in the variable a which, since $|\mathbf{x}^*[(z + b)^n]| \leq \|\mathbf{x}^*\|[|z| + |b|]^n$, will converge for $|a| \leq \frac{\rho}{4M}$ (noting here that even though a is restricted to be real, the power series in the variable a will converge for complex a in the range indicated). Since this power series is zero for all small real positive a, by the uniqueness Theorem 22 for power series all the coefficients in the series must be zero.

Because f is not a polynomial, there are arbitrarily large values of $n = m$ for which $\gamma_m \neq 0$, and for such m, $\mathbf{x}^*[(z + b)^m] = 0$. Using the binomial theorem, write

$$0 = \mathbf{x}^*[(z+b)^m] = \sum \binom{m}{j} b^{m-j} \mathbf{x}^*(z^j).$$

This expression is a polynomial in b which is zero for the infinite number of values of b in the interval $[0, M]$ and hence is identically zero, so that all the coefficients of this polynomial are zero and thus $\mathbf{x}^*(z^j) = 0$, for $j = 0, 1, \ldots, m$. Since m can be arbitrarily large, the continuous linear functional $\mathbf{x}^*$ is zero on all powers of z, hence on all polynomials, contradicting $\mathbf{x}^*(p) \neq 0$.   Q.E.D.

Note that the real part of the terms in the sum (5.2)

$$Re[c_n f(a_n(z + b_n))] = Re(c_n) Re(f(a_n(z + b_n)) - Im(c_n) Im(f(a_n(z + b_n)))$$

are each harmonic, since $Re(f) = U$ and $Im(f) = V$ are harmonic and
$Ref(a(z + b)) = U(ax + b, ay + b)$ has

$$\Delta U(ax + b, ay + b) = a^2 U_{xx}(ax + b, ay + b) + a^2 U_{yy}(ax + b, ay + b) = 0$$

and ditto for the second term. Hence, the real part of the function $h(z)$ of Eq. (5.2) is harmonic in the domain D and approximately equal to g on the boundary.

The restrictions on the coefficients a and b in the theorem can be easily modified, say for numerical purposes. This is a minor example of a much deeper truth: A theorem must be read to understand what it proves, a mathematical theorem is a collection of facts and techniques with only a few of the consequences actually stated in the theorem.

It is possible to take the coefficient b in the theorem to be a fixed point, not a variable quantity, see [2] and the exercises, but this point must be chosen so that f has a specific property, and this is less useful.

Corollary 6 *Theorem 24 and the Walsh-Lebesgue Theorem 23 are equivalent.*

Proof The proof of Theorem 24 shows that the Walsh-Lebesgue Theorem implies the results of the theorem.

Assume Theorem 24, and let g be a continuous real-value function on Γ and let $\epsilon > 0$ be given. By Theorem 24, there is a function $h(z)$ as in Eq. 5.2 with $|Re(h(z)) - g(z)| < \epsilon$ for all z on Γ. The parameters a_n and b_n in $h(z)$ have been chosen so that $|a_n(z + b_n)| < \frac{\rho}{2}$ and so the power series for each of the terms $f(a_n(z + b_n))$ converges uniformly for z on Γ. Set $M = 2 \sum_0^N |c_n|$. For each of the N terms $f(a_n(z + b_n))$, choose an index N_n with the property that $f(a_n(z + b_n))$ is approximated to within $\frac{\epsilon}{M}$ by the first N_n terms of its power series expansion, noting that any finite partial sum of the power series is a polynomial. If the power series for each term $f(a_n(z + b_n))$ is taken to have $\max(N_1, N_2, \ldots, N_N)$ terms, then each term is approximated to within $\frac{\epsilon}{M}$, and $Re(h(z))$ differs from this sum $p(z)$ of truncated terms, which is a polynomial, by less than $\sum_1^N |c_n| \frac{\epsilon}{M} < \epsilon$. And this polynomial $p(z)$ has $|Re(p(z)) - g(z)| < 2\epsilon$, which gives the Walsh-Lebesgue Theorem.   Q.E.D.

Example 16 This example shows that it is necessary to suppose that the function f in the above theorem not be a polynomial. Suppose, instead, that f is a polynomial of degree m. Let D be an bounded open set contained in $\{z : |z| < r\}$ with boundary Γ. Let $\mathcal{M}$ be the subspace of $C(\Gamma)$ spanned by the $\{1, z, z^2, \ldots, z^m, \bar{z}, \bar{z}^2, \ldots, \bar{z}^m\}$. Then any element of $\mathcal{M}$ has the form $p(z) + \overline{q(z)}$, $p(z)$ and $q(z)$ polynomials in $C(\Gamma)$ of degree less than or equal to m. Since $f(z)$ is assumed to be a polynomial of degree m, each term of the form $f(a(z+b))$ in the sum defining $h(z)$ in (5.2) is also a polynomial of degree m, and so $h(z)$ belongs to $\mathcal{M}$. Further, the structure of $\mathcal{M}$ shows that $\overline{h(z)}$ also belongs to $\mathcal{M}$ and therefore so does $Re(h(z)) = \frac{1}{2}[h(z) + \overline{h(z)}]$.

Next we show that $Re(z^{m+1})$ does not belong to $\mathcal{M}$. If it did, we could write $Re(z^{m+1}) = p(z) + \overline{q(z)}$, $p(z)$ and $q(z)$ polynomials of degree no more than m, and thus

$$Re(z^{m+1}) = Re(p(z)) + Re(q(z)) = Re(p(z) + q(z)) = Re(s(z))$$

with $s(z)$ a polynomial degree less than or equal to m. Then $Re(z^{m+1} - s(z))$ is harmonic, since $z^{m+1} - s(z)$ is a polynomial and so a power series converging for all z, and is zero on Γ so by the Maximum Principle is then zero in all of the open set D. In terms of real and imaginary parts

$$z^{m+1} - s(z) = 0 + iIm(z^{m+1} - s(z)) = 0 + iv(x, y).$$

Applying the Cauchy-Riemann equations in D, $v_x = v_y = 0$, and v itself must be a constant c_0. But then
$$z^{m+1} = s(z) + ic_0$$

holds in the open set D. In this set the above equation can be differentiated $m + 1$ times to obtain the contradiction $(m + 1)! = 0$.

Since $\mathcal{M}$ is finite dimensional, it is complete and so a closed subspace, and by the Hahn-Banach Theorem there is a continuous linear function $\mathbf{x}^*$ on $C(\Gamma)$ with $\mathbf{x}^*(\mathcal{M}) = 0$ and $\mathbf{x}^*(Re(z^{m+1})) = \delta \neq 0$. Finally let $g(z) = Re(z^{m+1})$ be the given continuous boundary condition function on Γ. For any $h(z)$ in $\mathcal{M}$,

$$\|Re(h(z)) - g(z)\| \geq \frac{|\mathbf{x}^*(Re(h) - g)|}{\|\mathbf{x}^*\|} = \frac{\delta}{\|\mathbf{x}^*\|} > 0$$

which shows that $g(z)$ cannot be approximated to an arbitrary degree of accuracy using sums of the form (24).

The solution of problems by means of approximating sums like those used in this chapter were first called The Complex Variable Boundary Element Method, see [3] and [4]. These methods used a functions of the form $(z - \beta)\log(z - \beta)$, the development of which depends on contour integration over Γ. For the use of a larger class of functions as in Theorem 24, see [2]. The use of power series has been a method for getting to the main results quickly. But using contour integrals displays the deep power of complex analysis, and the reader is encouraged to learn more about these methods.

5.1 Exercises

1. An infinite set A is countable if its elements can be indexed by the positive integers $Z = \{1, 2, \ldots, n, \ldots\}$, i.e. $A = \{a_1, a_2, \ldots, a_n, \ldots\}$. Another way of saying this is that there is a one-to-one function $\phi : Z \to A$ mapping Z onto A. (And since it is a one-to-one function there is an inverse function mapping A onto Z).

(a) Show that if there is a function $\phi : Z \to A$ mapping Z onto A, then there is a one-to-one function Ψ mapping Z onto A. Hint: Define Ψ inductively: First, $\Psi(1) = \phi(1)$, and set $n_1 = 1$. Next, let $n_2 > 1$ be the first integer with $\phi(n_2) \neq \phi(1)$ and let $\Psi(2) = \phi(n_2)$. Suppose that $\Psi(j)$ has been defined for $j = 1, 2, \ldots, p$; $\Psi(n_j) = \phi(n_j)$ with all the $\phi(n_j)$, distinct, $j = 1, \ldots, p$. How should $\Psi(p+1)$ be defined so as to complete its inductive definition?

(b) Let $Z \times Z$ denote the set of ordered pairs of positive integers: $Z \times Z = \{(i, j) : i = 1, 2, \ldots, j = 1, 2 \ldots\}$. Write $Z \times Z$ as an array, a matrix:

$$(1, 1), (1, 2), (1, 3), \ldots$$

$$(2, 1), (2, 2), (2, 3), \ldots$$

$$(3, 1), (3, 2), (3, 3), \ldots$$

$$\cdots$$

Define Ψ mapping Z onto $Z \times Z$ as follows (Draw the path of the values of Ψ on the array). $\Psi(1) = (1, 1)$, $\Psi(2) = (1, 2)$, $\Psi(3) = (2, 1)$, $\Psi(4) = (3, 1)$, $\Psi(5) = (2, 2)$, $\Psi(6) = (1, 3)$. What are the next few values of Ψ? From the diagram you can see that Ψ is onto. By part (a) there is a one-to-one map of Z onto $Z \times Z$ This is surprising because $Z \times Z$ is clearly "bigger", and for finite sets this is true, but infinite sets have interesting properties that are at first not intuitive.

(c) Suppose that we have a countable collection of sets A_n, each of which is countable or finite. Let A be the union of all these sets: $A = \cup_{n=1}^{\infty} A_n = $ the set of all x which belong to at least one of the sets A_n. Argue as in part (b) above to show that A is countable. (A can be finite only in very special cases, which are?)

(d) (The Cantor Diagonal Argument). Cantor's result was that the real numbers R are not countable. At that time, it was a revolutionary argument.

Define a map, using the tangent function, that is a one-to-one map of the interval $(0, 1)$ onto the real numbers R. This shows that R is countable exactly if $(0, 1)$ is countable. In Chap. 2, it was shown, as consequence of the completeness of the real numbers, that each real number in $(0, 1)$ has a corresponding convergent decimal expansion, $x = .a_1 a_2 \ldots a_n \ldots$, the numbers $a_1, a_2, \ldots$ each being one of the integers

$\{0, 1, 2, 3, 4, 5, 6, 7, 8, 9\}$. Given this, suppose that the numbers in $(0, 1)$ are count-able set. $\{x_1, x_2, \ldots\}$, with decimal expansions $x_j = .a_{j,1}a_{j,2}\ldots a_{j,n}\ldots$. Arrange these decimal coefficients in an array similar to what was done to show that the rational numbers were countable:

$$x_1 = .a_{1,1}a_{1,2}a_{1,3}, \ldots$$
$$x_2 = .a_{2,1}a_{2,2}a_{2,3}, \ldots$$
$$\ldots$$
$$x_n = .a_{n,1}a_{n,2}a_{n,3}, \ldots$$
$$\ldots$$

There is one possible ambiguity in a decimal expansion, an example of which is $\frac{1}{4} = .2500000\ldots = .249999999\ldots$. (Show this using the geometric series). To eliminate this, always use the expansion that terminates in zeros. Using the diagonal of this array of integers define a real number $y = .b_1b_2b_3\ldots$ in $(0, 1)$ as follows: For $j = 1, 2, \ldots$, set $b_j = 4$ if $a_{j,j} = 6$ and $b_j = 6$ if $a_{j,j} \neq 6$. The number y so constructed is not on the list of what was supposed to be a complete list of all the numbers in $(0, 1)$, under the assumption that $(0, 1)$ was countable. Why is that? This contradiction shows that $(0, 1)$ cannot be countable.

2. See [2].

(a) Suppose that the function f which is used in the approximating function $h(z) = \sum_1^N c_n f(a_n(z + b_n))$ of Theorem 24 has the property that $f^{(n)}(0) \neq 0$, for $n = 0$ and for all the derivatives $n = 1, 2, \ldots$. By modifying the proof of this theorem, show that the conclusion of the theorem is true using sums of the form $h(z) = \sum_0^N c_n f(a_n z)$ by using the fact that no coefficient $\frac{f^{(n)}(0)}{n!}$ is zero. In the next two parts of this exercise it will be shown that such a point z_0 exists.

(b) In Theorem 24, the function $f(z)$ has a power series expansion which con-verges for $|z| < \rho$, and the arguments $a_n(z + b_n)$ are restricted by bounds on a_n and b_n so that $w = a_n(z + b_n)$ satisfies $|w| < \frac{\rho}{2}$ for all z in D. It will simplify notation to assume here that the domain D is contained in the set of z with $|z| < \frac{\rho}{2}$. Recall, also, that f is not a polynomial. Choose $\delta > 0$ with $\frac{\rho}{2} < \delta < \rho$ so that $0 < \rho - \delta < \frac{\rho}{2}$. Consider the interval $[0, \delta]$. The goal now is to show that there is a point z_0 in this interval for which $f^{(n)}(z_0) \neq 0$ for $n = 0, 1, 2 \ldots$. Assume this is not the case. Let A_n be the set of all x in $[0, \delta]$ with $f^{(n)}(x) = 0$. Because it is assumed that there is not a point z_0 in the interval with the property described, it must be that each x in the interval belongs to at least one of the sets A_n, that is to say that $[0, \delta] = \cup_{n=0}^{\infty} A_n$. We now argue that each A_n is finite or contains no points. To show this, suppose that for some value of n, A_n is infinite. This being the case, there are points $x_1, x_2, \ldots$ in A_n. By compactness, there is a subsequence converging to x_0. The function $f(z)$ can be expanded in a power series in the variable $z - x_0$ (Why?). This function $g(z) = f(z - x_0)$ with $g^{(n)}(z) = f^{(n)}(z - x_0)$, has $g^{(n)}(z)$ zero on a sequence of points converging to zero, and by Theorem 22 $g^{(n)}(z)$ is zero in its circle of convergence and so in that circle $g(z)$ is a polynomial of degree not more than n. Why does this imply that this is also true for the function f, which contradicts the assumption that f is not a polynomial?

It has been shown that no A_n is infinite, and therefore $[0, \delta]$ is either countable or finite. This is a contradiction, since it implies there is a point z_0 in the uncountable set $[0, \delta]$ which belongs to none of the sets A_n and so for this point $f^{(n)}(z_0) \neq 0$ for all n.

3. For the functions $f(z) = e^z$, $f(z) = \cos(z)$, and $f(z) = \sin(z)$, find a point z_0 as in problem 2. Do this for some other functions. How about $f(z) = \frac{1}{1-z^2}$?

References

1. Ransford T (1995) Potential theory in the complex Plane. Cambridge University Press, Cambridge
2. Whitley R, Hromadka T II (2001) A general complex variable boundary element method, Numer Methods Partial Differ Equ 17:332–335
3. Hromadka T II, Lai C (1987) The complex variable boundary element method in engineering analysis, Springer, New York
4. Hromadka T II, Whitley R (1998) Advances in the complex variable boundary element method, Springer, New York

Chapter 6
The R^N Dirichlet Problem

Abstract In this chapter the notation is made simpler by considering harmonic functions which are complex valued functions and have real and imaginary parts satisfying the Laplace equation in N-variables. The Dirichlet problem is to find such a function of N variables, satisfying the Laplace equation in the interior of a domain D and being equal to a given continuous complex-valued function g on the boundary Γ of D. The case of interest in applications is $N=3$, but there is no essential difference in considering general N. In addition to specifying that the bounded open set D not have any holes, bubbles in three dimensions, which was enough in two dimensions, additional conditions are necessary in three or higher dimensions, and the relevant Poincaré condition is discussed. The proof uses the Hahn-Banach Theorem applied in the space $C(K)$ of complex-valued continuous functions defined on the boundary K of the domain. In the final result it is shown that there is an approximate solution using any one chosen function $f(z)$, which is not a polynomial and which has a power series convergent (about z = 0). This approximate solution is a finite sum of the form $c_n f(a^{(n)} \cdot x + b^n \cdot x + d_n)$ where c_n and d_n are complex numbers, $a^{(n)}$ and b^n are vectors in R^N with the same length, $\|a^{(n)}\| = \|b^{(n)}\|$, which are perpendicular $a^{(n)} \cdot b^n = 0$, and x is a point in R^N. The proof relies on a lemma which shows how to construct a harmonic function in N variables from a harmonic function in two variables, which will be applied to the real and imaginary parts of $f(z)$, and a representation theorem for the space of harmonic polynomials in N-variables which are homogeneous of degree m. Note that this result demonstrates the ability of a function of two variables, $f(z)$, to generate an approximate solution of the Dirichlet problem in N-variables.

Keywords Harmonic (in R^N) · Arcwise-connected · Poincaré truncated cone condition · harmonic polynomials homogeneous of degree m.

This chapter will extend the results of the previous chapter from R^2 to R^N. Problems in R^2 have, in the previous chapters, led to mathematical methods of great applicabil-

T. Hromadka and R. Whitley, *Foundations of the Complex Variable Boundary*   69
Element Method, SpringerBriefs in Applied Sciences and Technology,
DOI: 10.1007/978-3-319-05954-9_6, © The Author(s) 2014

ity, but realistic physical problems, say for the heat equation, are naturally modeled in R^3. For applications, take $N = 3$ in what follows.

However, the results for R^N are surprising in that, as in R^2, solutions to the Dirichlet problem in R^N can by approximated using a single non-polynomial power series (i.e., an analytic function), which is a function of the complex variable $z = x + iy$, so a function of only two real variables x and y. It is a worthwhile exercise to restate the results here for R^N to R^3 using the traditional three variables x, y, and z.

Recall that a function $U(\mathbf{x})$ on R^N, $\mathbf{x} = (x_1, x_2, \ldots, x_N)$, with continuous second partial derivatives, is harmonic if it satisfies:

$$\Delta U(\mathbf{x}) = \frac{\partial^2 U(\mathbf{x})}{\partial x_1^2} + \cdots + \frac{\partial^2 U(\mathbf{x})}{\partial x_N^2} = 0 \tag{6.1}$$

And we will also use the elegant way of writing a polynomial in N variables:

$$P(\mathbf{x}) = \sum_\alpha c_\alpha \mathbf{x}^\alpha \tag{6.2}$$

where $\alpha = (n_1, n_2, \ldots, n_N)$ is an N-tuple of non-negative integers, the sum is over a finite number of α since $P(\mathbf{x})$ is a polynomial, $\mathbf{x} = (x_1, x_2, \ldots, x_N)$ is an element of R^N, and $\mathbf{x}^\alpha = x_1^{n_1} x_2^{n_2} \cdots x_N^{n_N}$ (Write a few polynomials in three variables using this notation).

Recall that in Chap. 3 two important fact were established about polynomials in N variables:

$$P(\mathbf{x}) = \sum_m \sum_{|\alpha|=m} C_\alpha \mathbf{x}^\alpha = \sum_m P_m(\mathbf{x}) \tag{6.3}$$

is a unique expansion of $P(\mathbf{x})$ in terms of polynomials $P_m(\mathbf{x})$ which are homogeneous of degree m, i.e., for positive t we have $P_m(t\mathbf{x}) = t^m P(\mathbf{x})$. The terms in $P_m(\mathbf{x})$ are those where the sum of the powers of the variables x_i, $i = 1, 2, \ldots, N$ in each term is m. The other fact is that $P(\mathbf{x})$ is harmonic if and only if all the polynomials $P_m(\mathbf{x})$ are harmonic. These facts will be used to reduce a problem concerning harmonic polynomials to a problem concerning harmonic polynomials which in addition are homogeneous of degree m.

Previously, influenced by an interpretation based on the flow of heat, we have taken harmonic functions to be real-valued. To simplify notation it is convenient here to take harmonic functions to be complex-valued, so to say that $U(\mathbf{x}) = ReU(\mathbf{x}) + iImU(\mathbf{x})$ is harmonic means that its real and imaginary parts are harmonic:

$$\Delta U(\mathbf{x}) = \Delta ReU(\mathbf{x}) + i\Delta ImU(\mathbf{x})$$

and $\Delta U(\mathbf{x})$ will be zero exactly when that is true of both $ReU(\mathbf{x})$ and $ImU(\mathbf{x})$.

In order to generalize Theorem 24 from two to N dimensions we need a way of using a power series function of the variable $z = x+iy$, which is a harmonic function of the two real variables x and y (in the extended sense that its real and imaginary parts are both real-valued harmonic functions) to obtain harmonic functions of N variables. It will help to reconsider the case of two variables.

Lemma 15 *Let $\mathcal{H}_m(R^2)$ be the set of harmonic polynomials in two variables x_1 and x_2 which are homogeneous of degree m. Then $\mathcal{H}_m(R^2)$ is the complex vector space of polynomials with basis $(x_1 + ix_2)^m$ and $(x_1 - ix_2)^m$.*

$$\mathcal{H}_m(R^2) = sp\{(x_1 + ix_2)^m, (x_1 - ix_2)^m\} \tag{6.4}$$

Proof The proof is by induction on the integer m.

The result is true for $m = 0$, although in this special case $\mathcal{H}_0(R^2)$ is only one-dimensional, and is true for $m = 1$.

Suppose the result holds for all non-negative integers less than or equal to m and consider a function U belonging to $\mathcal{H}_{m+1}(R^2)$. The function $\frac{\partial U}{\partial x_1}$ belongs to $\mathcal{H}_m(R^2)$ (Why?) and so can be written as

$$\frac{\partial U}{\partial x_1} = a(x_1 + ix_2)^m + b(x_1 - ix_2)^m.$$

a and b constants. From this it follows that

$$U = \frac{a}{m+1}(x_1 + ix_2)^{m+1} + \frac{b}{m+1}(x_1 - ix_2)^{m+1} + q(x_2),$$

where $q(x_2)$ is a homogeneous polynomial of degree $m + 1$ in the single variable x_2, from which $g(x_2) = cx_2^{m+1}$. The function U being harmonic shows that $c = 0$. Q.E.D.

The notation in what follows will be neater if the inner product of two vectors in R^N, for example $\mathbf{a}$ and $\mathbf{x}$, is denoted by the dot product notation for R^N, i.e.,

$$\langle \mathbf{a}, \mathbf{x} \rangle = \mathbf{a} \cdot \mathbf{x}.$$

Using this notation the functions in Eq. (6.4) can be written in the form

$$(x_1 \pm ix_2)^m = (\mathbf{a} \cdot \mathbf{x} \pm i\mathbf{b} \cdot \mathbf{x})^m$$

with $\mathbf{a} = (1, 0)$ and $\mathbf{b} = (0, 1)$ vectors in R^2 which have the same length in the standard R^2 metric, $\|\mathbf{a}\|_2 = (\mathbf{a} \cdot \mathbf{a})^{\frac{1}{2}} = \|\mathbf{b}\|_2 = 1$, and are perpendicular: $\mathbf{a} \cdot \mathbf{b} = 0$.

Lemma 16 *Let h be an harmonic function of two variables defined on an open set D in R^2, and let $\mathbf{a}$ and $\mathbf{b}$ be two vectors in R^N, which are perpendicular, $\mathbf{a} \cdot \mathbf{b} = 0$ and have the same length, $\|\mathbf{a}\|_2 = \|\mathbf{b}\|_2$. Then*

$$H(\mathbf{x}) = h(\mathbf{a} \cdot \mathbf{x}, \mathbf{b} \cdot \mathbf{x})$$

is a harmonic function of $\mathbf{x}$ *in* R^N *for all points* $(\mathbf{a} \cdot \mathbf{x}, \mathbf{b} \cdot \mathbf{x})$ *in the domain D.*

Proof Compute the Laplacian

$$\Delta H(\mathbf{x}) = h_{11}(\mathbf{a} \cdot \mathbf{x}, \mathbf{b} \cdot \mathbf{x}) \sum_1^N a_j^2 + h_{22}(\mathbf{a} \cdot \mathbf{x}, \mathbf{b} \cdot \mathbf{x}) \sum_1^N b_j^2 + 2h_{12}(\mathbf{a} \cdot \mathbf{x}, \mathbf{b} \cdot \mathbf{x}) \sum_1^N a_j b_j$$

where the subscripts on h denote partial derivatives with respect to the first (1) and second (2) variable.   Q.E.D.

Parallel to the development for R^2, an R^N version of the Walsh–Lebesgue Theorem will be used, and this requires that the bounded open domain D, with closure K, satisfy two conditions. The condition for R^2 was that $R^2 - K$ have no holes, the condition in R^N is basically the same, that $R^N - K$ has no bubbles, i.e.: (1) $R^N - K$ is arcwise-connected. The second condition was, historically a surprise [1, p 285]; based on the result for R^2 it was thought that the first condition would be also be sufficient in R^N. However, Lebesgue gave an example which can be described in R^3 as follows: Let D be a flexible balloon about the origin and push a sharp spine into it which has the shape of the curve $y = e^{-1/x}$, $0 < x < 1$, rotated about the x-axis. For this surface there are continuous boundary conditions for which the Dirichlet problem has no solution. Roughly, there is a problem at the boundary point $(0, 0, 0)$ because the extreme closeness of the boundary values near zero interferes with the desired value at zero. There is a nice condition, sufficient for applications, namely that (2) The domain must satisfy the Poincaré exterior cone condition ([2, p 186], [3, p 232]). This condition states that for each point on the boundary of the domain, for simplicity say $(0, 0, 0)$ in R^3, there is a truncated cone, given by rotating the line $y = \epsilon x, 0 < x < \epsilon, \epsilon > 0$, about the x-axis, which has the endpoint $(0, 0, 0)$ the boundary point and the rest of the truncated cone lies in the exterior of the domain (You can see how this rules out the sharp spine of Lebesgue's example).

Theorem 25 *[2, p 232] Let D be an open bounded set in R^N with closure K, which with $R^N - K$ arcwise-connected and which satisfies the Poincaré truncated cone condition. Given a continuous function g on the boundary of K and $\epsilon > 0$, there is an harmonic polynomial $P(\mathbf{x})$ with*

$$|P(\mathbf{x}) - g(\mathbf{x})| < \epsilon \text{ for all } \mathbf{x} \text{ in the boundary of } K.$$

Theorem 26 *Let $\mathcal{H}_m(R^N)$ denote the space of harmonic polynomials in N variables which are homogeneous of degree m, and let $\mathcal{AB}_N$ be the collection of all pairs $(\mathbf{a}, \mathbf{b})$ of vectors in R^N which satisfy $\mathbf{a} \cdot \mathbf{b} = 0$ and $\|\mathbf{a}\|_2 = \|\mathbf{b}\|_2$. The space $\mathcal{H}_m^N$ is equal to the complex vector space spanned by functions of the form $(\mathbf{a} \cdot \mathbf{x} + i\mathbf{b} \cdot \mathbf{x})^m$ for all pairs $(\mathbf{a}, \mathbf{b})$ in $\mathcal{AB}_N$.*

Proof The proof will be by induction on the dimension $N \geq 2$, and then for that N a further induction on those m for which the theorem holds for the N under consideration.

It has been shown that the theorem holds for $N = 2$ and all $m = 0, 1, \ldots$. Also, for any N, the theorem is true for $m = 0$; and for $m = 1$ a polynomial homogeneous of degree one is in the span of $x_j = \frac{1}{2}[(x_j + ix_k) + (x_j - ix_k)]$, $k \neq j$.

To start the induction suppose that the theorem is true for $N \geq 2$ and for all m. Now consider $N + 1$, and noting that the theorem is true for $m = 0$ and $m = 1$, we suppose that it holds for some m and show that it then holds for $m + 1$ (for this $N + 1$).

Let $P(\mathbf{x})$ be in $\mathcal{H}_{m+1}(R^{N+1})$. Writing $P(\mathbf{x})$ as

$$\sum_\alpha d_\alpha x_1^{\alpha_1} x_2^{\alpha_2} \ldots x_{N+1}^{\alpha_{N+1}},$$

shows that the partial derivative of P with respect to the first variable x_1, $D_1 P$, is harmonic because $D_1 \Delta P = \Delta D_1 P$, and so belongs to $\mathcal{H}_m(R^{N+1})$. By the induction hypothesis $D_1 P$ is a finite sum

$$D_1 P(\mathbf{x}) = \sum c_j \left(\mathbf{a}^{(j)} \cdot \mathbf{x} + i \mathbf{b}^{(j)} \cdot \mathbf{x} \right)^m,$$

the pairs $\mathbf{a}^{(j)}$ and $\mathbf{b}^{(j)}$ each in $\mathcal{AB}_{N+1}$. Define a polynomial Q by

$$Q(\mathbf{x}) = \sum c_j^* \left(\mathbf{a}^{(j)} \cdot \mathbf{x} + i \mathbf{b}^{(j)} \cdot \mathbf{x} \right)^{m+1},$$

where

$$c_j^* = \frac{c_j}{(m+1)\left(a_1^{(j)} + i b_1^{(j)}\right)}$$

if $a_1^{(j)} + i b_1^{(j)} \neq 0$, and otherwise take $c_j^* = 0$. We have $D_1[P(\mathbf{x}) - Q(\mathbf{x})] = 0$, noting that when $c_j^* = 0$, the corresponding term in P has x_1 to the zero power and thus D_1 acting on that term is zero. The difference $P - Q$ belongs to $\mathcal{H}_{m+1}\left(R^{N+1}\right)$ and by construction has $D_1(P - Q) = 0$. Since $P - Q$ is an harmonic polynomial, homogeneous of degree $m + 1$, it has the form $\sum c_\alpha \mathbf{x}^\alpha$, the sum over the distinct terms with $\alpha_1 + \alpha_2 + \cdots + \alpha_{N+1} = m + 1$, and because $D_1(P - Q) = 0$, the variable x_1 occurs to the zeroth power in each of the linearly independent terms. Thus the sum can be written:

$$P(\mathbf{x}) - Q(\mathbf{x}) = \sum c_{(0,\alpha_2,\alpha_3,\ldots,\alpha_{N+1})} x_2^{\alpha_2} x_3^{\alpha_3} \ldots x_{N+1}^{\alpha_{N+1}}$$

this sum taken over $\alpha_2 + \alpha_3 + \cdots + \alpha_{N+1} = m + 1$. This shows that $P - Q$ is an harmonic polynomial in the N variables $x_2, x_3, \ldots, x_{N+1}$, homogeneous of degree $m + 1$, and as such by the induction hypothesis can be written as a linear combination of terms of the form

$$[(a_2, a_3, \ldots, a_{N+1}) \cdot (x_2, x_3, \ldots, x_{N+1}) + i(b_2, b_3, \ldots, b_{N+1}) \cdot (x_2, x_3, \ldots, x_{N+1})],$$

the pair $(a_2, a_3, \ldots, a_{N+1})$ and $(b_2, b_3, \ldots, b_{N+1})$ belonging to $\mathcal{AB}_N$, and these terms can also be written in the form:

$$[(0, a_2, \ldots, a_{N+1}) \cdot (x_1, x_2, \ldots, x_{N+1}) + i(0, b_2, \ldots, b_{N+1}) \cdot (x_1, x_2, \ldots, x_{N+1}),$$

the pairs $(0, a_2, \ldots, a_{N+1})$, $(0, b_2, \ldots, b_{N+1})$ belonging to $\mathcal{AB}_{N+1}$. Hence, the sum representing $P - Q$ belongs to $\mathcal{H}_{m+1}(R^{N+1})$, as does Q and therefore so does P, which completes the proof by induction.   Q.E.D.

If $\|\mathbf{a}\| = \|\mathbf{b}\| \neq 0$,

$$(\mathbf{a} \cdot \mathbf{x} + i\mathbf{b} \cdot \mathbf{x})^m = \|\mathbf{a}\|^m \left(\frac{\mathbf{a}}{\|\mathbf{a}\|} \cdot \mathbf{x} + i\frac{\mathbf{b}}{\|\mathbf{b}\|} \cdot \mathbf{x} \right)^m,$$

and so the representation Theorem 26 holds if the pairs $(\mathbf{a}, \mathbf{b})$ in $\mathcal{AB}_N$ are restricted to have $\|\mathbf{a}\| = \|\mathbf{b}\| = r$, for some $r > 0$.

Theorem 27 *Let D be a bounded open set in R^N, with closure K, $R^N - K$ arcwise-connected, and which satisfies the Poincaré exterior cone condition. Suppose that $f(z)$, not a polynomial, has a power series about zero converging for $|z| < \rho$. Then given g, a continuous function mapping the boundary Γ of K into the complex numbers and $\epsilon > 0$, there are a finite number of pairs $\left(\mathbf{a}^{(j)}, \mathbf{b}^{(j)}\right)$ in $\mathcal{AB}_N$, complex constants c_j and d_j, all for $j = 1, 2, \ldots, n$, with*

$$\left| g(x) - \sum_{1}^{n} c_j f \left(\mathbf{a}^{(j)} \cdot \mathbf{x} + \mathbf{b}^{(j)} \cdot \mathbf{x} + d_j \right) \right| \leq \epsilon. \tag{6.5}$$

Conditions need to be placed on the arguments of f so that they will be in the domain of f and contain an open interval about zero; for example, if $\mathbf{x}$ in the bounded set D has $\|\mathbf{x}\| \leq K_0$, then $\left\|\mathbf{a}^{(j)}\right\| = \left\|\mathbf{b}^{(j)}\right\| < \frac{\rho}{8K_0}$ and $|d_j| < \frac{\rho}{4}$ will easily be enough, as will any finite bounds in the case $\rho = \infty$.

Proof The proof will utilize the Banach space $C(\Gamma)$ of complex-valued continuous functions defined on the metric space Γ, the boundary of the domain D, taken with the supremum norm: $\|g\| = \sup\{|g(z)| : z \text{ in } \Gamma\}$. Note that an harmonic polynomial $P(\mathbf{x})$ belongs to $C(\Gamma)$, the normed linear space hypothesis that $\|P\| = 0$ implies that $P = 0$ following from The Maximum Principle.

Let M be the subspace spanned by functions of the form $f(\mathbf{a} \cdot \mathbf{x} + i\mathbf{b} \cdot \mathbf{x} + d)$, with size restrictions, as indicated in the statement of the theorem, so that $\mathbf{a} \cdot \mathbf{x} + i\mathbf{b} \cdot \mathbf{x} + d$ so restricted has absolute value less than or equal to $\frac{\rho}{2}$ and so lies in the circle of convergence. The theorem will be true if M is dense in $C(\Gamma)$. Assume that M is not dense; for this to be true it must be that there is a continuous function g with $d(g, M) > 0$. Under the hypotheses on the domain D, by Theorem 25 this implies that there is an harmonic polynomial $P(\mathbf{x})$ with $d(P, M) > 0$. By the Hahn–Banach

Theorem, there is a continuous linear functional $\mathbf{x}^*$ on $C(\Gamma)$ with $\mathbf{x}^*(P) \neq 0$, and $\mathbf{x}^*(M) = 0$.

The power series $f(z) = \sum e_j z^j$ converges uniformly on closed circles with radius less than ρ, that is in the norm of $C(\Gamma)$, so that

$$0 = \mathbf{x}^* \left(\sum e_j (\mathbf{a} \cdot \mathbf{x} + i\mathbf{b} \cdot \mathbf{x} + d)^j \right) = \sum e_j \mathbf{x}^* \left[(\mathbf{a} \cdot \mathbf{x} + i\mathbf{b} \cdot \mathbf{x} + d)^j \right].$$

This holds for all $(\mathbf{a}, \mathbf{b})$ in $\mathcal{AB}_N$, with $\|\mathbf{a}\| = \|\mathbf{b}\| < r$; the exact value of $r > 0$ is not important. Take $\lambda = \|\mathbf{a}\|$, and rewrite the above as

$$\sum e_j \lambda^j \mathbf{x}^* (\mathbf{a}' \cdot \mathbf{x} + i\mathbf{b}' \cdot \mathbf{x} + d')^j = 0$$

with $\lambda < r$, $|d'| < r'$, and $\|\mathbf{a}'\| = \|\mathbf{b}'\| = 1$, again the exact values of r and r' are not important. This gives a power series in λ whose coefficients

$$e_j \mathbf{x}^* (\mathbf{a}' \cdot \mathbf{x} + i\mathbf{b}' \cdot \mathbf{x} + d')^j$$

must be zero. Because f is not a polynomial, there are arbitrarily large integers n for which $e_n \neq 0$, and for such an index

$$\mathbf{x}^* \left[(\mathbf{a}' \cdot \mathbf{x} + i\mathbf{b}' \cdot \mathbf{x} + d')^n \right] = 0.$$

By the binomial theorem

$$\sum_{k=0}^{n} \binom{n}{k} (\mathbf{x}^* \left[(\mathbf{a}' \cdot \mathbf{x} + i\mathbf{b}' \cdot \mathbf{x})^k \right] (d')^{n-k} = 0.$$

The above is a polynomial of degree less than or equal to n in d', the equation holding for all small $|d'|$, with more than n zeros, and thus

$$\mathbf{x}^* \left(\left[\mathbf{a}' \cdot \mathbf{x} + i\mathbf{b}' \cdot \mathbf{x} \right]^k \right) = 0.$$

holds for $k = 0, 1, 2, \ldots, n$. Since n can be taken to be arbitrarily large, apply Theorem 26 to see $\mathbf{x}^*(Q) = 0$ for all harmonic homogeneous polynomials Q of degrees $k = 0, 1, \ldots$, and, since P can be written as a finite sum of harmonic homogeneous polynomials, $\mathbf{x}^*(P) = 0$, contrary to assumption.   Q.E.D.

6.1 Exercises

1. State and prove the R^N version of Corollary 6 in Chap. 5.
2. In what sense are Theorem 25 and 5 generalizations of the Weierstrass Approximation Theorem? Compare the three types of domains.

References

1. Kellogg O (1953) Foundations of potential theory. Springer, Berlin (Reprint Dover, New York, 1953)
2. Armitage D, Gardiner S (2000) Classical potential theory. Springer, London
3. Axler S, Bourdon P, Ramey W (2001) Harmonic function theory. Springer, New York

Bibliography

1. Conway J (1990) A course in functional analysis, 2nd edn. Springer, New York
2. Gardiner S (1995) Harmonic approximation. Cambridge University Press, Cambridge
3. Schechter M (2001) Principles of functional analysis, 2nd edn. American Mathematical Society, Providence
4. Smith K (1971) Primer of modern analysis. Bogden and Quigley, New York

T. Hromadka and R. Whitley, *Foundations of the Complex Variable Boundary Element Method*, SpringerBriefs in Applied Sciences and Technology, DOI: 10.1007/978-3-319-05954-9, © The Author(s) 2014

Index

T. Hromadka and R. Whitley, *Foundations of the Complex Variable Boundary Element Method*, SpringerBriefs in Applied Sciences and Technology, DOI: 10.1007/978-3-319-05954-9, © The Author(s) 2014